Quantum Physics for Beginners

Discover Quantum Mechanics from Fundamental Particles to the Unexplained Mysteries of the Universe

Robert Meis

Copyright ©2024 All rights reserved

The content contained within this book may not be reproduced, duplicated or transmitted without direct written permission from the author or the publisher.

Under no circumstances will any blame or legal responsibility be held against the publisher, or author, for any damages, reparation, or monetary loss due to the information contained within this book, either directly or indirectly.

Legal Notice: This book is copyright protected. This book is only for personal use. You cannot amend, distribute, sell, use, quote or paraphrase any part, or the content within this book, without the consent of the author or publisher.

Disclaimer Notice: Please note the information contained within this document is for educational and entertainment purposes only. All effort has been executed to present accurate, up to date, and reliable, complete information. Readers acknowledge that the author is not engaging in the rendering of legal, financial, medical or professional advice.

By reading this document, the reader agrees that under no circumstances is the author responsible for any losses, direct or indirect, which are incurred as a result of the use of the information contained within this document, including, but not limited to, — errors, omissions, or inaccuracies.

Table of Contents

The Birth of Quantum Theory _____ 5

Schrödinger's Cat and the Wave Function _____ 19

The Uncertainty Principle _____ 29

Quantum Entanglement _____ 39

The Copenhagen Interpretation _____ 47

Quantum Cosmology and the Multiverse _____ 54

Quantum Field Theory _____ 69

Pilot-Wave Theory _____ 79

Unsolved Mysteries in Quantum Physics _____ 83

New Frontiers _____ 103

Recommended Resources _____ 119

Afterword _____ 125

The Birth of Quantum Theory

Max Planck and the Absolute Blackbody

At the dawn of the 20th century, physics was largely dominated by the classical theories of mechanics and electromagnetism. These theories, developed by Isaac Newton, James Clerk Maxwell, and others, provided a comprehensive framework for understanding the behavior of objects and forces on macroscopic scales. However, as experimental techniques improved and physicists began investigating phenomena on the atomic scale, they encountered problems that classical physics could not explain. One of the most significant of these problems was the issue of blackbody radiation, a dilemma that ultimately led to the birth of quantum theory and the groundbreaking work of Max Planck.

Blackbody radiation refers to the electromagnetic radiation emitted by an idealized object, called a "blackbody," which absorbs all incident radiation without reflecting any. The concept of a blackbody is an abstraction, but it serves as a useful model for understanding how objects radiate energy. In the late 19th century, physicists sought to understand the relationship between the temperature of a blackbody and the spectrum of radiation it emitted. Classical physics, specifically the laws of thermodynamics and electromagnetism, were used to derive a theoretical explanation for blackbody radiation. However, this led to a significant contradiction known as the "ultraviolet catastrophe."

The ultraviolet catastrophe emerged from a key prediction made by classical physics, which suggested that as the wavelength of radiation emitted by a blackbody decreased, the intensity of the radiation should increase without bound. According to this prediction, blackbodies should emit an infinite amount of energy at ultraviolet and shorter wavelengths, a result that was clearly at odds with experimental observations. In reality, the intensity of radiation at shorter wavelengths (such as ultraviolet) peaked and then diminished, rather than continuing to increase indefinitely. This stark inconsistency between theory and experiment represented a profound crisis in classical physics.

The roots of the ultraviolet catastrophe can be traced back to the Rayleigh-Jeans law, a formula derived using classical mechanics to describe the distribution of energy in blackbody radiation. The law worked well at longer wavelengths, such as infrared radiation, but completely failed at shorter wavelengths. As the wavelength decreased, the Rayleigh-Jeans law predicted that the intensity of radiation would increase dramatically, leading to an infinite energy output in the ultraviolet region—hence the term "ultraviolet catastrophe."

Physicists at the time were puzzled by this paradox. Attempts to modify classical theories to account for the observed behavior of blackbody radiation were unsuccessful. It became clear that a new approach was needed, one that could explain why blackbodies did not emit infinite amounts of energy at high frequencies and short wavelengths.

The breakthrough came in 1900 when Max Planck, a German physicist, proposed a radical new idea. Planck hypothesized that energy is not emitted continuously, as classical physics suggested, but rather in discrete packets, which he called "quanta." In other words, Planck proposed that energy is quantized, meaning it can only be emitted or absorbed in specific, discrete amounts rather than in a smooth, continuous flow. This idea marked the beginning of quantum theory, although Planck himself initially viewed his hypothesis as a mathematical convenience rather than a fundamental change to physics.

Planck's hypothesis was based on his work to derive a formula that accurately described the observed spectrum of blackbody radiation. By assuming that the energy of oscillators (the microscopic particles responsible for emitting radiation) was quantized, Planck was able to derive a formula that fit experimental data perfectly. This formula, now known as Planck's law, described the distribution of energy in blackbody radiation as a function of wavelength and temperature. Unlike the Rayleigh-Jeans law, Planck's law predicted that the intensity of radiation would peak at a certain wavelength and then decrease at shorter wavelengths, which matched experimental results.

The key to Planck's success was his assumption that the energy of each oscillator could only take on discrete values, rather than any arbitrary value as classical physics would allow. This was a revolutionary idea because it challenged the classical view that energy could vary continuously.

At first, Planck regarded his hypothesis as an ad hoc solution to a specific problem, but its implications were far-reaching. The introduction of quantized energy was a major departure from classical physics and marked the first step toward the development of quantum mechanics. In fact, Planck's constant became one of the fundamental constants in all of quantum theory, representing the scale at which quantum effects become significant.

The success of Planck's theory in explaining blackbody radiation paved the way for future developments in quantum physics. It resolved the ultraviolet catastrophe by showing that at higher frequencies, the energy of radiation was quantized, which prevented the infinite energy output predicted by classical theories. Planck's work also set the stage for the subsequent discoveries of quantum phenomena, including Albert Einstein's explanation of the photoelectric effect in 1905, which provided further evidence for the quantization of energy and helped establish quantum theory as a new framework for understanding the physical world.

Einstein and the Photoelectric Effect

In 1905, Albert Einstein published a paper that would revolutionize our understanding of light and set the stage for the development of quantum mechanics. His work on the photoelectric effect provided compelling evidence for the particle nature of light, a concept that challenged the classical wave theory of light and introduced the notion that light, which had been traditionally understood as a wave, could also behave as a particle. This discovery would ultimately lead to the development of quantum theory and would later earn Einstein the Nobel Prize in Physics in 1921.

The photoelectric effect refers to the phenomenon in which light shining on the surface of a material, typically a metal, causes the emission of electrons from that material. This effect had been observed in experiments as early as the 19th century, but classical physics could not adequately explain the underlying mechanism. According to the wave theory of light, which was widely accepted at the time, light was thought to be a continuous electromagnetic wave that could transfer energy to electrons over time, eventually causing them to be emitted from the material. However, experimental results from the photoelectric effect contradicted this explanation, and a new theory was needed.

One of the key problems that classical wave theory faced in explaining the photoelectric effect was the relationship between the frequency of light and the energy of the emitted electrons. Classical theory predicted that the energy of the emitted electrons should depend on the intensity (or brightness) of the light, not its frequency. In other words, according to classical physics, the brighter the light, the more energy the electrons should have, regardless of the light's color (which corresponds to its frequency). However, experiments showed that this was not the case. The energy of the emitted electrons depended only on the frequency of the light, not its intensity. Moreover, no electrons were emitted if the frequency of the light was below a certain threshold, no matter how bright the light was. This was a puzzling result that classical physics could not explain.

Einstein's explanation of the photoelectric effect was both simple and radical. He proposed that light is not just a continuous wave but is made up of discrete packets of energy, which he called "quanta" (later referred to as photons). Each photon carries a specific amount of energy that is proportional to the frequency of the light. In this framework, light of higher frequency (such as ultraviolet light) is composed of higher-energy photons, while light of lower frequency (such as red light) is made up of lower-energy photons.

When light shines on a material, Einstein suggested that it is not the intensity or brightness of the light that determines whether electrons are emitted, but the energy of the individual photons. If a photon has enough energy, it can transfer that energy to an electron in the material, causing the electron to escape from the surface. If the photon does not have enough energy, the electron cannot escape, no matter how many photons are present or how intense the light is. This explanation successfully accounted for the experimental results of the photoelectric effect, which showed that the energy of the emitted electrons depends on the frequency of the light, not its intensity.

The threshold frequency observed in the experiments could now be explained by Einstein's theory. Below a certain frequency, the photons do not have enough energy to overcome the work function of the material—the minimum energy required to release an electron from the surface. Even if the light is very intense, if the photons do not have the required energy (i.e., if the light's frequency is too low), no electrons will be emitted. Conversely, light of a higher frequency, even if it is very dim, will cause electrons to be emitted because the photons carry enough energy to knock the electrons loose from the material.

This was a dramatic departure from the classical wave theory of light, which had dominated physics since the work of James Clerk Maxwell in the 19th century. Maxwell's equations had successfully described light as an electromagnetic wave, and this wave theory had been highly successful in explaining phenomena such as reflection, refraction, diffraction, and interference. However, the photoelectric effect could not be explained by treating light solely as a wave, and Einstein's particle theory of light offered a solution.

The concept of photons, or light quanta, was a key step in the development of quantum mechanics. It showed that electromagnetic radiation has a dual nature—it can behave both as a wave and as a particle, depending on the context. This wave-particle duality is a fundamental aspect of quantum physics and applies not only to light but also to other particles, such as electrons, which can exhibit both wave-like and particle-like properties. The idea that light can exist as discrete packets of energy, or photons, was revolutionary and marked a significant shift in our understanding of the nature of light and energy.

In addition, the discovery of the photoelectric effect has had practical applications in technology. Devices such as photovoltaic cells, which convert sunlight into electricity, operate based on the principles of the photoelectric effect. This technology forms the basis of solar panels and has played a key role in the development of renewable energy sources.

Wave-Particle Duality

Historically, the concepts of "waves" and "particles" were distinct and separate in classical physics. Waves, like sound waves or water waves, were understood as continuous disturbances that propagate through space and time. They could interfere with each other, exhibit diffraction (spreading out when passing through a narrow opening), and were characterized by properties such as wavelength and frequency. Particles, on the other hand, were considered discrete entities with mass and definite positions. They followed predictable trajectories governed by Newton's laws of motion and interacted with other particles through collisions or forces.

The idea that something could behave both like a wave and a particle would have been unthinkable in the classical framework. However, this traditional view began to break down as scientists investigated the behavior of light and matter on extremely small scales. The study of light, in particular, revealed early hints of the dual nature of physical entities.

The debate over the nature of light stretches back centuries. In the 17th century, Isaac Newton proposed that light was composed of particles, which he called "corpuscles." At the same time, Dutch physicist Christiaan Huygens argued that light was a wave, citing phenomena such as diffraction and interference as evidence for the wave theory. For many years, both models coexisted, with each being applied to explain different aspects of light's behavior. By the 19th century, however, the wave theory gained dominance, largely due to the work of James Clerk

Maxwell, who developed a set of equations (Maxwell's equations) that described light as an electromagnetic wave.

Yet, as experiments in the early 20th century revealed more about the behavior of light and particles, it became clear that neither the wave nor the particle theory of light was sufficient on its own. While Einstein's explanation of the photoelectric effect provided strong evidence for the particle nature of light, other experiments showed that light also exhibited wave-like properties.

One of the most striking demonstrations of light's wave-like behavior is the double-slit experiment. In this classic experiment, light is passed through two closely spaced slits, and the resulting pattern is observed on a screen behind the slits. Instead of forming two distinct bands on the screen, as one would expect if light were composed purely of particles, an interference pattern emerges. This pattern consists of alternating bright and dark bands, characteristic of wave interference, where waves from the two slits overlap and either reinforce or cancel each other out.

The interference pattern observed in the double-slit experiment suggests that light is a wave. However, when individual photons are sent through the slits one at a time, the same interference pattern eventually emerges, even though each photon should behave like a particle and pass through one slit or the other. This strange result implies that even individual particles of light "interfere" with themselves, as if each photon somehow travels through both slits simultaneously and behaves like a wave before being detected as a particle on the screen.

The double-slit experiment does not only apply to light. It can also be performed with electrons, atoms, and even larger molecules, all of which show similar wave-like behavior when passed through the slits. When electrons, for example, are fired one at a time through the slits, an interference pattern eventually appears, just as with light. This suggests that matter, like light, can exhibit both wave-like and particle-like properties. The wave aspect of particles is described by what is known as the wave function, a mathematical function that provides information about the probability of finding a particle in a particular location.

The wave-particle duality of electrons was first proposed by French physicist Louis de Broglie in 1924. De Broglie suggested that just as light can behave as both a wave and a particle, particles of matter, such as electrons, can also exhibit wave-like properties. He proposed that the wavelength associated with a particle is inversely proportional to its momentum, a relation now known as the de Broglie wavelength. This hypothesis was confirmed experimentally a few years later when electrons were shown to produce diffraction patterns, just like waves.

Wave-particle duality poses a fundamental challenge to classical intuition. How can something be both a wave and a particle? The answer lies in the probabilistic nature of quantum mechanics. According to quantum theory, particles do not have definite positions or velocities until they are measured. Instead, their behavior is described by a wave function, which encodes the probability of finding the particle at a particular location. Before measurement, a particle exists in a superposition of all possible states, meaning it has both wave-like and particle-like characteristics. Once measured, the wave function collapses, and the particle assumes a definite position or momentum, behaving like a classical particle.

Wave-particle duality also raises philosophical questions about the nature of reality. If particles can behave like waves, and waves can behave like particles, what exactly are they? There is no definitive answer to this question to this day.

Bohr's Model of the Atom

Prior to Bohr's contributions, the prevailing model of the atom was based on classical physics, particularly the work of Ernest Rutherford, who had proposed a nuclear model of the atom just a few years earlier. Rutherford's model depicted the atom as a miniature solar system, with a small, positively charged nucleus at the center and electrons orbiting around it, much like planets around the Sun. While this model was revolutionary in many ways, it left significant gaps in explaining the behavior of electrons and the stability of atoms, especially when viewed in the light of classical electromagnetic theory.

One of the central challenges with Rutherford's model was its inability to explain why electrons, which were assumed to move in circular or elliptical orbits around the nucleus, did not gradually spiral into the nucleus due to the emission of electromagnetic radiation. According to classical physics, an accelerating charged particle, such as an electron in orbit, should continuously emit radiation and lose energy. As the electron lost energy, it would be expected to spiral closer to the nucleus, eventually causing the atom to collapse. Clearly, this did not happen in reality, as atoms are stable structures. Furthermore, Rutherford's model could not explain why atoms emitted or absorbed light in discrete wavelengths, as observed in the spectra of elements like hydrogen.

Bohr set out to address these issues by introducing a new model of the atom that incorporated ideas from quantum theory, which was just beginning to take shape at the time. His model was based on several key postulates, the most important of which was that electrons could only occupy certain discrete energy levels, or "orbits," around the nucleus. Unlike the continuous orbits proposed by classical mechanics, Bohr's orbits were quantized, meaning that electrons could only exist in specific, allowed orbits with fixed energies. These allowed orbits corresponded to particular energy levels, and as long as an electron remained in a given orbit, it would not radiate energy and thus would not spiral into the nucleus.

This quantization of energy levels was a radical departure from classical physics, but it provided a straightforward explanation for the stability of atoms. In Bohr's model, the electron could occupy one of these quantized orbits without radiating energy, and thus the atom remained stable. Moreover, Bohr proposed that electrons could transition between these orbits by absorbing or emitting a photon of light, with the energy of the photon corresponding to the difference between the energy levels of the two orbits. This explained why atoms emitted or absorbed light in discrete wavelengths, as the energy of the emitted or absorbed photon was directly related to the difference in energy between the initial and final states of the electron.

Bohr's model was particularly successful in explaining the spectrum of hydrogen, the simplest atom with just one electron orbiting a single proton in the nucleus. The spectral lines of hydrogen, which had been observed for many years, appeared as distinct, discrete lines rather than a continuous spectrum. These lines corresponded to the emission or absorption of light at specific wavelengths, and Bohr's model provided a clear explanation for this. According to his theory, the electron in a hydrogen atom could only occupy certain allowed energy levels, and when it jumped from a higher energy level to a lower one, it emitted a photon of light with an energy corresponding to the difference between the two levels. This energy difference determined the wavelength of the emitted light, resulting in the observed spectral lines.

The success of Bohr's model in explaining the hydrogen spectrum was a major triumph for quantum theory, and it provided strong evidence for the idea that energy levels in atoms were quantized. However, while Bohr's model worked remarkably well for hydrogen, it encountered difficulties when applied to more complex atoms with multiple electrons. The model also could not fully explain the fine structure of spectral lines, which indicated the presence of additional energy levels or sublevels. These shortcomings would eventually be addressed by the more complete and sophisticated quantum mechanical models developed in the following years, particularly the development of wave mechanics by Erwin Schrödinger and matrix mechanics by Werner Heisenberg.

Despite its limitations, Bohr's model was a crucial stepping stone in the development of modern quantum mechanics. It introduced the idea that classical physics could not fully describe the behavior of particles on the atomic scale and that new, quantum principles were required to explain the structure of atoms and the interactions between their components.

Schrödinger's Cat and the Wave Function

Waves are one of the most fundamental concepts in physics, describing the propagation of energy through space and matter. Whether we are talking about the ripples on a pond, sound waves traveling through the air, or light waves crossing vast distances in space, the behavior of waves can be described using a common set of principles and mathematical tools. Understanding the nature of waves is essential in the study of quantum mechanics, as many quantum phenomena, including the behavior of particles like electrons, are governed by wave-like properties. In fact, one of the key discoveries of quantum mechanics is that particles at the atomic and subatomic levels often behave more like waves than classical objects.

A wave can be thought of as a disturbance or oscillation that travels through space and transfers energy from one location to another without transferring matter. Waves can be classified into different types based on how they propagate. For example, mechanical waves, such as sound waves or water waves, require a medium (like air or water) to travel through, while electromagnetic waves, like light, can travel through a vacuum. Waves can also be longitudinal or transverse, depending on the direction in which the oscillations occur relative to the direction of the wave's propagation.

In a longitudinal wave, the oscillations of the particles in the medium are parallel to the direction the wave is traveling. Sound waves are a good example of this type of wave: as the wave moves through the air, the air molecules vibrate back and forth along the direction of the wave's movement, creating regions of compression and rarefaction. In contrast, transverse waves are characterized by oscillations that are perpendicular to the direction of wave travel. Light waves and water waves are examples of transverse waves, where the oscillations (either of electric and magnetic fields, in the case of light, or of water molecules) occur at right angles to the wave's direction of propagation.

The amplitude of a wave is the maximum displacement of a point on the wave from its equilibrium position. For example, in a water wave, the amplitude is the height of the wave crest above the still water level. In a sound wave, the amplitude corresponds to the intensity or loudness of the sound, while in a light wave, the amplitude is related to the brightness of the light. The amplitude of a wave is often associated with the energy it carries: the larger the amplitude, the more energy the wave transports.

Finally, the velocity of a wave is the speed at which the wave propagates through a medium. In general, the speed of a wave depends on both the type of wave and the medium through which it is traveling. For example, sound waves travel faster through denser materials like water or steel than through air, because the molecules in denser materials are closer together and can transmit the oscillations more efficiently. The speed of light in a vacuum, on the other hand, is a universal constant. In other media, such as glass or water, light travels more slowly due to interactions with the material.

One of the most important equations in wave theory is the wave equation, which describes how waves propagate through space and time.

One of the most important equations in wave theory is the wave equation, which describes how waves propagate through space and time.

If the waves are in phase, meaning their crests and troughs align, they will reinforce each other, leading to constructive interference and a wave with greater amplitude. If the waves are out of phase, meaning the crest of one wave aligns with the trough of another, they will cancel each other out, resulting in destructive interference. Interference patterns are a key feature in many quantum experiments, including the famous double-slit experiment, which demonstrated the wave-like behavior of particles such as electrons.

Probability and the Wave Function

Unlike classical physics, which is deterministic and allows for precise predictions about the behavior of objects, quantum mechanics deals with probabilities and uncertainties. This shift in how we understand physical systems is most clearly seen through the idea of the wave function, a mathematical tool that encodes everything we can know about a quantum system. The wave function doesn't tell us where a particle is or what its momentum is with absolute certainty; instead, it provides the probabilities of finding the particle in a particular place or state when a measurement is made.

This wave-like behavior is captured by the wave function, which represents the state of a particle or system of particles. The wave function, typically denoted by the symbol "psi," is not something we can observe directly, but it is the key to understanding how particles behave at the quantum level.

The wave function can be thought of as a sort of "probability wave." It is spread out over space, meaning that a particle described by this wave function doesn't exist in a single, definite location. Rather, the particle has a range of possible positions it might occupy, each with a different probability. The wave function itself doesn't give the probability directly, but when we square its value, we obtain what is called the probability density. This probability density tells us how likely we are to find the particle at a particular point in space if we were to make a measurement.

In a classical system, if we know an object's position and velocity at a given moment, we can predict its future behavior with precision. For example, if we know the position and speed of a baseball, we can calculate where it will land. In quantum mechanics, the wave function prevents such certainty. Instead of knowing exactly where a particle is, we only have access to the probabilities of where it might be found. These probabilities are spread out over a range of locations. The particle could be here, there, or somewhere in between, and we can only calculate the likelihood of each outcome.

This probabilistic interpretation of the wave function was formalized by Max Born in the 1920s. Born suggested that the wave function should not be interpreted as representing the physical shape or path of a particle, but rather as a tool for calculating the probability of where a particle might be found. This interpretation—known as the Born rule—revolutionized the way physicists thought about quantum systems and marked a significant departure from the certainty of classical mechanics.

The probabilistic nature of quantum mechanics also introduces the concept of uncertainty. The wave function inherently contains information about both the position and momentum of a particle, but these two properties are not independent of one another. In fact, there's a fundamental limit to how precisely we can know both at the same time, a principle known as the Heisenberg uncertainty principle. The more precisely we know a particle's position, the less precisely we can know its momentum, and vice versa. This uncertainty is not due to imperfections in our measurements or instruments; it is a fundamental property of nature itself.

Consider a particle confined to a small region of space, such as an electron in an atom. The electron's wave function is spread out over the region surrounding the nucleus, and the shape of this wave function gives us the probability of finding the electron at different distances from the nucleus. Even though the electron is most likely to be found close to the nucleus, there is always a small but non-zero probability of finding it further away. The wave function provides a complete description of this situation, telling us the likelihood of all possible outcomes.

But quantum mechanics doesn't stop at position and momentum. The wave function can describe probabilities for all sorts of physical properties, including energy levels, spin, and even the polarization of photons. When we solve Schrödinger's equation—a key equation in quantum mechanics—we are often solving for the wave function that describes a system in terms of these variables. The result is a probability distribution for each property, giving us a comprehensive picture of all the possible states the system could be in, along with their respective probabilities.

Importantly, the probabilistic interpretation of the wave function doesn't mean that reality is random or chaotic. The outcomes of measurements in quantum mechanics are governed by strict mathematical rules. The wave function evolves smoothly and deterministically according to Schrödinger's equation, but the results of measurements—when we probe the system—are probabilistic. This dual nature of quantum mechanics, where the evolution of the wave function is deterministic but the outcomes of measurements are probabilistic, is one of the most fascinating aspects of the theory.

Superposition Principle

The superposition principle is fundamental ideas in quantum mechanics. It introduces a concept that defies our classical intuition: that particles, such as electrons or photons, can exist in multiple states simultaneously until they are measured or observed.

In classical physics, when we think of objects—whether they are as large as planets or as small as grains of sand—

we assume that they have definite properties at all times. For example, a planet has a specific location in its orbit and moves with a certain speed. If we were to measure its position or velocity at any given moment, we could pinpoint those values with precision. Classical objects exist in a single, well-defined state at any given time.

Quantum particles, however, do not behave in this straightforward way. In the quantum world, particles do not have definite properties until they are measured. Instead, they can exist in a combination of multiple possible states all at once. This is the essence of the superposition principle: a particle can be in a superposition of different states, meaning it doesn't have a single, fixed property but rather a mixture of all possible outcomes. It is only when a measurement is made that the particle "chooses" one of these possible states and behaves like a classical object with definite properties.

To better grasp this idea, let's consider an electron in a quantum system. According to classical physics, an electron could be thought of as orbiting a nucleus in a definite path, much like a planet orbits the sun. However, quantum mechanics tells us that the electron doesn't follow a precise path or exist in a definite position until we measure it. Instead, the electron's state is described by a wave function, which contains information about all the possible positions the electron might occupy. The electron exists in a superposition of all these possible positions at once, with each possible position having a certain probability.

The notion of superposition is often illustrated with the famous thought experiment known as Schrödinger's cat. This thought experiment, devised by physicist Erwin Schrödinger in the 1930s, involves a hypothetical cat placed in a sealed box along with a mechanism that has a 50-50 chance of killing the cat, depending on the outcome of a quantum event, such as the decay of a radioactive atom. According to quantum mechanics, until the box is opened and the cat is observed, the cat is in a superposition of being both alive and dead simultaneously. Only when we observe the cat does the superposition collapse into one of the two possible outcomes—either the cat is alive or it is dead. This paradox highlights the strange implications of the superposition principle in the quantum world, where objects can exist in multiple, seemingly contradictory states until measured.

A more concrete example of superposition can be found in the behavior of light particles, or photons, when they pass through a beam splitter in an experiment designed to test quantum behavior. When a photon encounters a beam splitter, it can either pass through the splitter or be reflected by it. According to classical physics, the photon must take one path or the other. But in quantum mechanics, the photon enters a superposition, where it is simultaneously traveling both paths. It's only when we measure the photon at a detector that it collapses into one specific path, revealing itself as having taken one route or the other.

This phenomenon is not limited to photons. Electrons, atoms, and even larger molecules can all exist in superposition, displaying wave-like behavior as they simultaneously occupy multiple states. In the famous double-slit experiment, for example, when individual electrons are fired one at a time through two narrow slits, they produce an interference pattern on a screen behind the slits, a pattern that indicates the electrons are behaving as if they passed through both slits at the same time. This is a clear demonstration of superposition: the electron exists in a superposition of going through both slits until a measurement is made, at which point it collapses into a definite state of having passed through one slit or the other.

The superposition principle also has profound implications for the concept of measurement in quantum mechanics. The act of measuring a quantum system forces the system to collapse from its superposition into a single, definite state. Before the measurement, the system exists in a mixture of possibilities, but once we observe it, we obtain a specific outcome. This raises deep philosophical questions about the nature of reality in the quantum realm. Is the quantum system in a superposition until we measure it, or does it "choose" a state before we observe it? The superposition principle seems to suggest that reality at the quantum level is not as fixed or objective as we are used to thinking in classical terms.

The superposition principle is also the key to understanding many of the technologies that are being developed based on quantum mechanics. One of the most promising applications of superposition is in the field of quantum computing. Classical computers process information using bits, which can exist in one of two states: 0 or 1. Quantum computers, on the other hand, use quantum bits or "qubits," which can exist in a superposition of both 0 and 1 simultaneously. This means that quantum computers have the potential to perform many calculations at once, vastly increasing their computational power compared to classical computers. Superposition, along with another quantum property called entanglement, allows quantum computers to solve certain types of problems much more efficiently than classical computers ever could.

The superposition principle also plays a role in other quantum technologies, such as quantum cryptography and quantum sensing. Quantum cryptography takes advantage of the superposition of quantum states to ensure secure communication channels that are resistant to eavesdropping. Quantum sensors, meanwhile, use superposition to make extremely precise measurements of physical quantities, such as magnetic fields or gravitational waves, with applications ranging from medical imaging to navigation systems.

Despite the strange and non-intuitive nature of superposition, it has been experimentally confirmed many times over. Physicists have observed superposition in electrons, photons, atoms, and even large molecules consisting of hundreds of atoms. These experiments confirm that superposition is not merely a theoretical construct but a fundamental feature of the quantum world.

The Uncertainty Principle

Heisenberg's Uncertainty Principle is one of the most famous and profound concepts in quantum mechanics. It highlights a fundamental limit to what we can know about a particle's state, and it marks a radical departure from the certainty offered by classical physics. In essence, the uncertainty principle tells us that certain pairs of physical properties, such as position and momentum, cannot both be known to arbitrary precision at the same time. The more precisely we know one property, the less precisely we can know the other. This principle is not a result of imperfections in our measurement instruments, but rather a fundamental aspect of nature itself.

To understand why this limitation exists, we need to first consider how particles behave in the quantum world. In classical mechanics, the state of a particle is fully determined by its position and momentum at any given time. If we know where a particle is and how fast it is moving, we can use the laws of classical physics to predict its future behavior with great accuracy. This determinism is central to classical physics, where knowing the initial conditions of a system allows us to calculate its entire future.

Quantum mechanics, however, reveals a different picture. At the atomic and subatomic scales, particles like electrons do not behave as solid, point-like objects with precise locations and velocities. Instead, their behavior is described by a wave function, which encodes a range of possible positions and momenta. The wave function spreads out over space, indicating that the particle doesn't have a single, well-defined position, but rather a probability distribution of where it might be found. This wave-like nature of particles lies at the heart of Heisenberg's Uncertainty Principle.

The wave function describes both the particle's position and its momentum, but here's where the problem arises: position and momentum are what physicists call conjugate variables, meaning they are related in such a way that they cannot both be known precisely at the same time. If we try to measure the particle's position very accurately, the wave function becomes sharply localized, representing a precise location. However, this causes the wave function to spread out in terms of momentum, meaning we lose information about the particle's momentum. Conversely, if we measure the particle's momentum with high precision, the wave function becomes spread out in terms of position, and we can no longer pinpoint the particle's exact location.

This trade-off is not due to a flaw in our measuring instruments, but is instead an intrinsic property of quantum systems. It is built into the fabric of quantum mechanics itself. No matter how advanced our technology becomes, the uncertainty principle sets a fundamental limit on how much we can know about a particle's state. The more precisely we try to determine one aspect of a particle's state, the less certain we become about the other. This introduces a level of uncertainty that is completely foreign to classical mechanics, where both position and momentum can, in principle, be known exactly.

One way to think about the uncertainty principle is to consider how we measure a particle. Imagine trying to measure the position of an electron. To do this, we need to interact with it in some way—perhaps by shining light on it. But light itself is made of photons, and when a photon interacts with the electron, it disturbs the electron's momentum. If we use light with a very short wavelength (high-energy photons) to get a more precise measurement of the electron's position, we end up imparting more energy to the electron, making its momentum more uncertain. On the other hand, if we use light with a longer wavelength (lower-energy photons) to minimize the disturbance to the electron's momentum, we lose precision in determining its position. This unavoidable disturbance is one way to understand why the uncertainty principle arises.

Heisenberg's Uncertainty Principle has profound implications for how we think about the nature of reality. In the classical world, we are used to the idea that objects have definite properties—things exist in specific places, moving at specific speeds, and we can measure these properties with sufficient precision. In the quantum world, this certainty dissolves. Particles no longer have definite, fixed properties until we measure them. Instead, they exist in a cloud of probabilities, where their position and momentum are only known with a certain degree of uncertainty. This probabilistic nature of quantum mechanics is unsettling for many, as it implies that the universe at its most fundamental level is governed by uncertainty and chance, rather than by strict cause and effect.

Another significant consequence of the uncertainty principle is that it forbids particles from having zero energy. In classical mechanics, if a particle is at rest, we would say that its position is fixed and its momentum is zero. However, in quantum mechanics, the uncertainty principle prevents us from knowing both the position and momentum exactly. This implies that even a particle that appears to be at rest must still possess some residual motion, known as "zero-point energy." This has important implications for fields such as quantum field theory and cosmology, where the uncertainty principle plays a role in phenomena like vacuum fluctuations, the energy of empty space, and even the early dynamics of the universe.

The uncertainty principle also plays a crucial role in explaining the stability of atoms. In classical physics, an electron orbiting a nucleus should, according to the laws of electromagnetism, continuously emit radiation and eventually spiral into the nucleus, causing the atom to collapse. But atoms are stable, and quantum mechanics provides the explanation. The uncertainty principle prevents the electron from being confined to an exact position at the nucleus. If the electron were confined too closely to the nucleus, its momentum would become highly uncertain and very large, causing it to "spread out" and move away from the nucleus. This balance between position and momentum, governed by the uncertainty principle, helps maintain the stability of atoms.

Regardless of the interpretation, the uncertainty principle sets a clear limit on the precision with which we can predict the behavior of quantum systems. This limitation has practical implications as well. For example, in the realm of technology, the uncertainty principle imposes constraints on how precisely we can manipulate or measure quantum systems. In fields like quantum computing and quantum cryptography, engineers and scientists must work within the bounds set by the uncertainty principle, which limits the precision of operations involving quantum bits (qubits) and other quantum systems.

The uncertainty principle also affects our understanding of energy and time. Just as there is an uncertainty between position and momentum, there is a similar relationship between energy and time. This means that, over very short time scales, there can be significant fluctuations in energy. This phenomenon gives rise to what is known as quantum fluctuations, which have far-reaching consequences in areas such as particle physics and cosmology. Quantum fluctuations are thought to play a role in the creation of virtual particles and in the early stages of the universe's development, particularly in theories related to the Big Bang and the inflationary model of cosmology.

On a more philosophical level, the uncertainty principle raises questions about the nature of knowledge itself. In classical physics, the universe is often compared to a giant clockwork machine, where every part moves in predictable ways, and all we need to do is uncover the details. But in quantum mechanics, the uncertainty principle suggests that there are limits to what we can know about the universe, even in principle. This does not imply that the universe is chaotic or random, but rather that the universe operates according to a set of rules that are fundamentally different from those we experience in our macroscopic, everyday world.

The uncertainty principle also invites us to reconsider the relationship between the observer and the observed. In classical physics, measurement is seen as a passive act—merely recording information about a pre-existing system. In quantum mechanics, however, measurement is an active process that influences the system being measured. This interplay between the observer and the system has led some to suggest that consciousness or the act of observation plays a role in shaping reality, although this idea remains a topic of intense debate within both the physics and philosophy communities.

Thought Experiments

One of the most famous thought experiments associated with the uncertainty principle is Heisenberg's own "gamma-ray microscope" experiment. This thought experiment was designed by Werner Heisenberg himself to illustrate the intrinsic limits of measurement imposed by his uncertainty principle. In this scenario, we imagine using a powerful microscope that employs gamma rays to observe the position of an electron. The goal is to pinpoint the electron's position as precisely as possible. However, because gamma rays have very short wavelengths, they carry a lot of energy. When one of these high-energy photons interacts with the electron to allow us to "see" it, the collision transfers momentum to the electron, thereby disturbing its motion.

In this thought experiment, the key idea is that by using higher-energy photons (such as gamma rays), we can measure the electron's position with greater precision because shorter wavelengths allow for finer resolution.

However, the downside is that the interaction with the photon causes a significant disturbance in the electron's momentum. Conversely, if we use lower-energy photons with longer wavelengths to reduce the disturbance to the electron's momentum, we lose precision in measuring its position. This trade-off between the precision of position and momentum measurements lies at the heart of the uncertainty principle. It is not that we lack the technology to make more precise measurements, but rather that nature itself prevents us from knowing both position and momentum simultaneously with arbitrary precision. Heisenberg's gamma-ray microscope thought experiment neatly captures the unavoidable consequences of trying to measure a quantum system.

Another thought experiment that emphasizes quantum uncertainty is the "quantum eraser" experiment, a variation of the double-slit experiment. In this setup, an additional layer of complexity is added by allowing the measurement of the particle's path to be "erased" after it has passed through the slits but before it reaches the detector. The results of this experiment show that if the "which-path" information is erased, the interference pattern reappears, even though the particle has already passed through the slits. This thought experiment challenges our classical understanding of time and causality, showing that the mere availability of certain information (like the particle's path) impacts the outcome. It reinforces the idea that quantum systems remain in a superposition of possibilities until a measurement is made and that the information gained during the measurement fundamentally affects the outcome.

Practical Application of the Uncertainty Principle

Quantum cryptography leverages the principles of quantum mechanics to create secure communication channels that are fundamentally resistant to eavesdropping. One of the most well-known protocols in quantum cryptography is Quantum Key Distribution (QKD), with the most famous example being the BB84 protocol, developed by Charles Bennett and Gilles Brassard in 1984.

The basic idea behind quantum cryptography is that any attempt to measure or eavesdrop on a quantum system inevitably disturbs the system due to the uncertainty principle. In QKD, information is encoded in the quantum states of particles, typically photons, and transmitted between two parties. Because of the uncertainty principle, if an eavesdropper tries to intercept and measure the quantum states, they will inevitably introduce detectable disturbances into the system. These disturbances will alert the legitimate parties to the presence of the eavesdropper, allowing them to discard the compromised data and ensure that their communication remains secure.

In more detail, QKD relies on the fact that quantum states cannot be measured without altering them. When two parties—commonly referred to as Alice and Bob—exchange information using quantum states, any attempt by an eavesdropper (Eve) to intercept and measure the transmitted particles will inevitably disturb their quantum properties. According to the uncertainty principle, Eve cannot gain perfect knowledge of the quantum states without introducing detectable errors into the transmission. As a result, QKD allows Alice and Bob to detect any intrusion and take appropriate steps to ensure the integrity of their communication.

Quantum cryptography has the potential to provide an unprecedented level of security for digital communications, particularly in an era where classical encryption methods are becoming increasingly vulnerable to sophisticated attacks. Traditional encryption methods rely on the computational difficulty of certain mathematical problems, such as factoring large numbers, to provide security. However, advances in quantum computing threaten to break these classical encryption algorithms, rendering them obsolete. In contrast, quantum cryptography, grounded in the uncertainty principle and the laws of quantum mechanics, offers a level of security that cannot be undermined by advances in computing power.

While the uncertainty principle itself does not directly govern the computational power of quantum computers, it influences how quantum systems behave, and it places limits on the precision with which we can manipulate and measure qubits. In particular, any measurement made on a qubit introduces some level of uncertainty and disturbance, which must be carefully managed to preserve the integrity of quantum information. This is one of the challenges in developing practical quantum computers, as maintaining coherence and minimizing error rates in qubits is essential for the reliable execution of quantum algorithms.

In addition to quantum cryptography and quantum computing, the uncertainty principle has practical applications in fields such as metrology, the science of measurement. The principle places limits on the precision with which certain measurements can be made, but it has also led to the development of technologies that take advantage of quantum uncertainty to improve measurement accuracy. For example, quantum sensors are being developed that can achieve levels of precision far beyond those possible with classical sensors. These quantum sensors use entangled particles or squeezed states to minimize uncertainties in certain variables, allowing for ultra-precise measurements of quantities such as time, gravitational fields, and magnetic fields. Applications of these quantum sensors range from improving the accuracy of atomic clocks to detecting gravitational waves and advancing medical imaging technologies.

In cosmology, the uncertainty principle is also thought to play a role in the early universe. Quantum fluctuations, small variations in energy that arise due to the uncertainty principle, may have contributed to the formation of the large-scale structure of the universe. These tiny fluctuations in the density of matter, amplified by cosmic inflation, are believed to have led to the formation of galaxies, stars, and other cosmic structures. Thus, the uncertainty principle has implications not only for microscopic systems but also for the largest structures in the universe.

Quantum Entanglement

Quantum entanglement describes a special type of connection between particles that can occur when they interact in certain ways, leading to correlations between their properties, even when they are separated by vast distances. Albert Einstein famously referred to quantum entanglement as "spooky action at a distance" because of the way it seems to violate the idea that information or influence cannot travel faster than the speed of light.

At its core, quantum entanglement occurs when two or more particles become linked in such a way that the state of one particle is dependent on the state of the other, no matter how far apart they are. When particles are entangled, measuring a property (such as the spin, polarization, or momentum) of one particle instantly determines the corresponding property of the other, even if the particles are on opposite sides of the universe. This instantaneous connection between entangled particles is what makes quantum entanglement so mysterious and counterintuitive.

When particles become entangled, their individual wave functions become linked, or "entangled," into a single, combined wave function that describes the entire system. As a result, the properties of the entangled particles are no longer independent of each other. Instead, they are correlated, meaning that the state of one particle is directly tied to the state of the other, regardless of the distance between them.

For example, imagine two particles that are entangled in such a way that their spins are always opposite—if one particle has an "up" spin, the other must have a "down" spin, and vice versa. According to quantum mechanics, before any measurement is made, both particles exist in a superposition of both spin states. It is only when we measure the spin of one particle that the wave function collapses, and the spin of both particles becomes definite. If we measure the first particle and find that it has an "up" spin, the second particle will instantly have a "down" spin, no matter how far away it is.

The key point is that this connection between the particles is not mediated by any signal or force traveling through space; it happens instantaneously. This instantaneous correlation appears to conflict with Einstein's theory of relativity, which holds that no information or influence can travel faster than the speed of light. However, quantum mechanics does not allow for faster-than-light communication between entangled particles. Even though the measurement of one particle affects the state of the other, this effect cannot be used to transmit information between distant observers in a way that would violate the principles of relativity. This aspect of entanglement is one of the reasons why it is so puzzling and continues to be the subject of intense study and debate.

Quantum entanglement also plays a role in developing new technologies like quantum teleportation, which involves the transfer of quantum information (such as the state of a particle) from one location to another without physically moving the particle itself. While this may sound like science fiction, quantum teleportation has already been demonstrated in laboratory settings, and it is a critical component of future quantum communication networks.

The discovery of quantum entanglement has not only expanded our understanding of the quantum world but also raised deep philosophical questions about the nature of reality. Entanglement challenges the classical notion of locality, which holds that objects are only directly influenced by their immediate surroundings. In the quantum world, particles can be connected across vast distances, leading some physicists and philosophers to rethink the very fabric of space and time. The implications of entanglement suggest that our classical, intuitive understanding of the universe may be incomplete, and that reality at the quantum level is far more interconnected and mysterious than we ever imagined.

Bell's Inequality

Bell's Theorem is one of the most significant developments in the history of quantum mechanics. It fundamentally changed how we understand the nature of reality, particularly challenging the classical concept of "local realism." This idea, which had been central to physics for centuries, assumes that objects have definite properties independently of observation (realism) and that these properties are influenced only by their immediate surroundings, with no faster-than-light interactions (locality). Bell's Theorem, however, showed that if quantum mechanics is correct, then either locality or realism—or both—must be abandoned. This result has profound implications for our understanding of the universe and has been experimentally confirmed, leaving classical notions of reality forever altered.

To grasp the importance of Bell's Theorem, we need to go back to the debate over the completeness of quantum mechanics. In the 1930s, Albert Einstein, Boris Podolsky, and Nathan Rosen published a famous paper now known as the "EPR paradox." Their argument was based on what they saw as a troubling feature of quantum mechanics: entanglement. According to quantum theory, two particles can become entangled in such a way that the state of one particle is directly linked to the state of the other, no matter how far apart they are. If one particle is measured, its entangled partner's state becomes immediately determined, even if the particles are light-years apart. This phenomenon seemed to suggest some kind of faster-than-light communication between the particles, which would violate Einstein's theory of relativity.

Einstein was deeply uncomfortable with this implication, famously referring to it as "spooky action at a distance." He and his colleagues argued that quantum mechanics must be incomplete. They believed that there must be hidden variables—unknown factors that quantum mechanics had yet to account for—that could explain the seemingly strange behavior of entangled particles. According to the EPR argument, these hidden variables would restore "local realism," the idea that particles have definite properties (realism) and that these properties are not influenced by distant objects (locality). Essentially, Einstein and his colleagues believed that quantum mechanics provided only a partial description of reality and that a deeper, more complete theory could resolve these paradoxes without violating locality.

For decades, this debate over the completeness of quantum mechanics remained largely theoretical. Then, in 1964, the Irish physicist John Bell published a groundbreaking paper that provided a way to test whether local realism was a valid concept or if quantum mechanics' predictions were truly correct. Bell devised a mathematical inequality—now known as Bell's inequality—that could be tested experimentally. The inequality was based on the assumption that local hidden variable theories, like those Einstein had proposed, should place certain limits on the correlations between measurements of entangled particles. If the correlations between entangled particles exceeded these limits, Bell's theorem suggested that local realism could not be correct, and the universe would have to operate in a way that defies classical intuition.

Bell's inequality provided a concrete way to settle the debate between quantum mechanics and local realism through experiments. According to Bell's theorem, if quantum mechanics is correct, the correlations between entangled particles would violate Bell's inequality, showing that no local hidden variable theory could explain the results. This would imply that either locality or realism (or both) had to be abandoned.

The first experimental tests of Bell's theorem were conducted in the 1970s and 1980s, most famously by the French physicist Alain Aspect and his colleagues. In these experiments, pairs of entangled photons were generated and sent in opposite directions to detectors, where measurements were made on the photons' polarizations. The results showed that the correlations between the photons' polarizations violated Bell's inequality, exactly as quantum mechanics had predicted. In other words, the experimental data ruled out any local hidden variable theory that could account for the behavior of the entangled particles. The results of these experiments provided strong evidence that the universe does not operate according to local realism.

One potential issue in early experiments was the so-called "locality loophole," which suggests that some hidden influence, traveling slower than light but still fast enough to affect the experiment, might have caused the observed correlations. Another issue was the "detection loophole," which arises if not all entangled particles are detected, potentially biasing the results.

In 2015, a series of "loophole-free" Bell tests were conducted, designed to close both the locality and detection loopholes simultaneously. One of the most notable of these experiments was performed by a team of researchers at Delft University of Technology in the Netherlands. In this experiment, pairs of entangled electrons were created and separated by more than a kilometer, with their spins measured at two distant locations. The experimenters ensured that the measurement events were space-like separated, meaning that no signal traveling at the speed of light could have influenced the outcome of both measurements. Moreover, they achieved high enough detection efficiencies to close the detection loophole as well.

The results of the Delft experiment, along with similar experiments conducted around the same time by other groups, provided the most convincing evidence yet for quantum entanglement. Once again, the correlations between the entangled particles violated Bell's inequality, confirming the predictions of quantum mechanics and ruling out any local hidden variable explanations. These experiments showed beyond a reasonable doubt that entanglement is real and that it cannot be explained by any classical theory based on local realism.

The experimental confirmation of quantum entanglement has had profound implications for both the foundations of physics and the development of new technologies. Entanglement is not just an abstract curiosity but a resource that can be harnessed for practical applications, such as quantum cryptography and quantum computing. In quantum cryptography, entangled particles are used to create secure communication channels that are immune to eavesdropping. This is because any attempt to intercept or measure the entangled particles would disturb the system, alerting the communicating parties to the presence of an eavesdropper. Quantum key distribution, based on entanglement, has already been demonstrated in real-world scenarios.

One of the most recent and ambitious applications of quantum entanglement is the development of a "quantum internet." In such a network, entangled particles would be distributed across vast distances, allowing for the instantaneous sharing of quantum information. This would enable new forms of communication, computation, and sensing that are impossible with classical technology.

Several experiments have already demonstrated the ability to distribute entangled particles over long distances, including via satellite-based quantum communication, marking the first steps toward a global quantum network.

What does this mean for our understanding of reality? Essentially, Bell's theorem forces us to reconsider how we think about the fundamental nature of the universe. One possibility is that the principle of locality is violated, meaning that entangled particles can influence each other instantly, regardless of the distance between them. This would imply some form of non-locality, where information or correlations can be transmitted faster than the speed of light. While this seems to contradict Einstein's theory of relativity, it's important to note that quantum entanglement does not allow for faster-than-light communication in the traditional sense. While the state of one particle instantly influences the state of its entangled partner, no usable information is transmitted in this process, so relativity's prohibition on faster-than-light signaling remains intact.

The Copenhagen Interpretation

The Copenhagen Interpretation is one of the most well-known and widely taught interpretations of quantum mechanics. Developed primarily in the 1920s by Danish physicist Niels Bohr and German physicist Werner Heisenberg, it provides a philosophical framework for understanding the strange and counterintuitive nature of the quantum world. While quantum mechanics itself is a mathematical theory that makes accurate predictions about the behavior of particles at the atomic and subatomic levels, the Copenhagen Interpretation addresses the question of what quantum mechanics means—how we should understand the results and implications of these predictions in terms of reality.

At its core, the Copenhagen Interpretation suggests that the act of measurement plays a central role in determining the state of a quantum system. Before a measurement is made, quantum systems are described by a wave function, which encodes all the possible states that the system can occupy. The wave function provides a complete description of the system in terms of probabilities, but these probabilities do not correspond to actual physical states until an observation or measurement is made. In other words, according to the Copenhagen Interpretation, a quantum system does not exist in a definite state before measurement; it exists in a superposition of multiple possible states.

One of the key ideas of the Copenhagen Interpretation is the notion of "wave function collapse." The wave function is a mathematical object that describes the range of possible outcomes for a quantum system. For example, an electron might have a probability of being found in different places around an atom, but before we measure its position, the electron exists in a superposition of all these possibilities. The wave function captures this superposition. However, when we perform a measurement—such as determining the electron's position—the wave function "collapses" to a single outcome, and the electron is observed in a specific location. After the collapse, the system no longer exists in a superposition; instead, it has a definite state based on the measurement result.

This leads to one of the most challenging aspects of the Copenhagen Interpretation: the idea that reality at the quantum level is not determined until a measurement is made. Unlike classical physics, where objects are assumed to have definite properties whether we measure them or not, quantum mechanics suggests that particles do not have definite properties—like position, momentum, or spin—until they are observed. Before measurement, these properties exist as probabilities, but they are not "real" in the classical sense. The measurement process itself brings the physical properties of the system into existence, which fundamentally changes the way we think about reality.

The Copenhagen Interpretation also avoids the need for "hidden variables"—theoretical constructs that would explain quantum behavior in terms of underlying, unobservable factors that determine the outcomes of measurements. In the early days of quantum mechanics, physicists like Albert Einstein were uncomfortable with the idea that quantum mechanics was inherently probabilistic and incomplete. Einstein famously remarked, "God does not play dice," expressing his belief that there must be hidden variables that could restore a deterministic understanding of reality. However, the Copenhagen Interpretation rejects the idea of hidden variables, arguing instead that quantum mechanics is complete as it stands, and that the probabilistic nature of the theory reflects a fundamental feature of reality.

The Copenhagen Interpretation is often seen as the "standard" or "textbook" interpretation of quantum mechanics, and it has been highly influential in shaping the way physicists think about the quantum world. However, it is not without its critics. Some physicists are uncomfortable with the idea that reality depends on measurement, and they argue that the Copenhagen Interpretation does not provide a fully satisfactory explanation of quantum phenomena.

Despite these alternative interpretations, the Copenhagen Interpretation remains a powerful and widely accepted way of thinking about quantum mechanics. It emphasizes the central role of probability, measurement, and uncertainty in the quantum world, and it challenges our classical intuitions about the nature of reality. By embracing the idea that quantum systems do not have definite properties until they are observed, the Copenhagen Interpretation provides a framework for understanding the counterintuitive behavior of particles at the smallest scales. It may not offer a fully deterministic view of the universe, but it has proven to be an extraordinarily successful interpretation for making accurate predictions about the behavior of quantum systems, and it continues to be a cornerstone of modern physics.

Criticism

While the Copenhagen Interpretation successfully describes and predicts the behavior of quantum systems, its implications for the nature of reality, measurement, and observation have sparked deep philosophical debates. These criticisms have led some physicists and philosophers to seek alternative interpretations of quantum mechanics, each attempting to address the perceived shortcomings of the Copenhagen view.

One of the central tenets of the Copenhagen Interpretation is that quantum systems exist in a superposition of possible states until they are measured. According to this view, the act of measurement collapses the wave function, and the system assumes a definite state—such as the position or momentum of a particle—only at the moment of observation. Before measurement, these properties are described probabilistically, with no definite reality. This raises the question of what "reality" means at the quantum level, as particles do not have fixed, well-defined properties until they are observed. For many critics, this idea is profoundly unsettling because it seems to suggest that reality is dependent on observation, challenging the classical notion of an objective, observer-independent world.

One of the most vocal critics of the Copenhagen Interpretation was Albert Einstein, who famously rejected the idea that quantum mechanics provided a complete description of reality. Einstein believed that there must be "hidden variables"—unknown factors that quantum mechanics does not account for—that would restore a deterministic understanding of the world. He was particularly troubled by the interpretation's apparent denial of realism, the idea that physical objects have definite properties, whether or not they are being observed. Einstein's discomfort with the Copenhagen Interpretation was famously encapsulated in his remark, "I like to think the moon is there even if I am not looking at it." This criticism underscores Einstein's belief that the universe should have an objective reality that exists independently of human observers.

Einstein, along with his colleagues Boris Podolsky and Nathan Rosen, also raised concerns about the Copenhagen Interpretation in the form of the EPR paradox, published in 1935. The EPR paper argued that quantum mechanics, as interpreted by the Copenhagen school, led to seemingly absurd consequences, such as "spooky action at a distance." In particular, the EPR paradox focused on the phenomenon of quantum entanglement, where two particles that have interacted in the past remain correlated even when separated by large distances. According to quantum mechanics, measuring the state of one particle instantly determines the state of the other, regardless of the distance between them. This appears to violate the principle of locality, which holds that objects can only be influenced by their immediate surroundings. Einstein and his colleagues argued that this paradox suggested that quantum mechanics was incomplete and that a deeper, more realistic theory was needed to resolve these issues.

Another common criticism of the Copenhagen Interpretation revolves around its treatment of measurement and the so-called "measurement problem." In the Copenhagen view, measurement plays a special role in quantum mechanics, as it is the process that causes the wave function to collapse and brings a quantum system into a definite state. However, this raises a difficult question: what exactly constitutes a "measurement"? Is it the involvement of a conscious observer that triggers the collapse, or can any interaction with a macroscopic system (such as a measuring device) count as a measurement? The Copenhagen Interpretation is famously vague on this point, leading to confusion about the boundary between the quantum and classical worlds. If a particle's wave function can remain in superposition until a measurement is made, at what point does the system transition from a quantum superposition to a definite classical state?

This ambiguity in the Copenhagen Interpretation was famously highlighted by the thought experiment known as "Schrödinger's cat," devised by physicist Erwin Schrödinger in 1935. According to the Copenhagen Interpretation, until the box is opened and the system is observed, the cat exists in a superposition of both alive and dead states. This paradox raises the question of whether macroscopic objects, like a cat, can also exist in superposition, and at what point the wave function collapse occurs. Schrödinger's thought experiment was intended as a critique of the Copenhagen Interpretation's reliance on the observer and measurement, illustrating how strange and seemingly absurd the implications of the theory could be when extended to everyday objects.

Another criticism of the Copenhagen Interpretation comes from its rejection of determinism. In classical physics, if we know the initial conditions of a system—such as the position and velocity of a particle—we can predict its future behavior with certainty. Quantum mechanics, as interpreted by the Copenhagen school, replaces this determinism with a probabilistic framework. The wave function provides the probabilities of different outcomes, but the exact outcome of a quantum measurement cannot be predicted with certainty. For some physicists, this probabilistic nature of quantum mechanics is unsatisfying because it suggests that the universe is not governed by strict cause and effect. Critics argue that this indeterminism is philosophically troubling and may point to a gap in our understanding of the underlying mechanics of quantum phenomena.

In response to these criticisms, several alternative interpretations of quantum mechanics have been proposed. One of the most prominent alternatives is the "many-worlds interpretation," first put forward by physicist Hugh Everett.

Quantum Cosmology and the Multiverse

Quantum cosmology is the branch of theoretical physics that seeks to apply the principles of quantum mechanics to the universe as a whole. While quantum mechanics is usually associated with subatomic particles and small-scale phenomena, quantum cosmology explores how these same principles might govern the very structure and evolution of the cosmos on the largest possible scales. This field attempts to bridge the gap between quantum mechanics, which explains the behavior of particles at the smallest scales, and general relativity, which describes the gravitational forces that shape stars, galaxies, and the universe as a whole.

One of the main challenges in quantum cosmology is reconciling the fundamentally different frameworks of quantum mechanics and general relativity. These two theories have both been incredibly successful in their respective domains, but they operate on different principles. Quantum mechanics is based on probabilistic events and uncertainties, while general relativity is a deterministic theory that treats gravity as a geometric property of spacetime. While general relativity works well for describing the large-scale structure of the universe, it breaks down when applied to extremely small scales, such as those that existed at the moment of the Big Bang or inside black holes. These are precisely the situations where quantum mechanics becomes important, suggesting that a quantum theory of gravity is needed to fully understand the early universe and the nature of spacetime.

One of the central ideas in quantum cosmology is that the universe itself might be subject to quantum effects. Just as particles can exist in superpositions of states, with multiple possible outcomes before measurement, some theories of quantum cosmology propose that the universe, too, might have originated in a superposition of different possible states. According to these theories, the universe could have begun as a quantum fluctuation in a "quantum vacuum," a state with no classical matter or energy but filled with quantum fields fluctuating at a subatomic level. These fluctuations could have given rise to the Big Bang, setting the universe in motion.

This idea draws on the concept of quantum fluctuations, which occur due to the uncertainty inherent in quantum mechanics. Even in a vacuum, quantum fields are never truly at rest; instead, they are constantly fluctuating. These tiny fluctuations can create pairs of particles and antiparticles, which typically annihilate each other almost immediately. However, under the right conditions—such as those that might have existed in the very early universe—these fluctuations could become significant enough to trigger the birth of a universe. This idea suggests that the Big Bang might not have been a singular, one-time event, but rather a quantum process that could, in principle, give rise to multiple universes, each emerging from its own quantum fluctuation.

One of the key theories that connects quantum mechanics and cosmology is the theory of inflation. Inflation proposes that the universe underwent a period of extremely rapid expansion in the first fraction of a second after the Big Bang. This inflationary phase smoothed out any irregularities in the early universe and set the stage for the formation of galaxies and other large-scale structures. What makes inflation particularly interesting from a quantum perspective is that it suggests quantum fluctuations that occurred during this early period were stretched across the universe, seeding the tiny variations in density that eventually grew into galaxies, stars, and planets.

Quantum cosmology also raises the possibility that our universe might be just one of many. Some theories suggest that quantum mechanics could allow for the existence of a "multiverse"—a collection of parallel universes that each have their own distinct properties and physical laws. In this view, our universe is just one branch in a vast multiverse that emerged from a quantum event. The idea of a multiverse is closely related to the Many-Worlds Interpretation of quantum mechanics, which holds that all possible outcomes of a quantum event occur in separate, non-interacting branches of reality. In a cosmological context, this means that there could be an infinite number of universes, each with its own unique history and physical characteristics.

Another intriguing area of quantum cosmology involves black holes. Black holes are regions of space where gravity is so strong that nothing, not even light, can escape. General relativity predicts that at the center of a black hole lies a "singularity," a point of infinite density where the laws of physics as we know them break down. However, quantum mechanics suggests that something more complex may be happening at the singularity. The interplay between quantum mechanics and general relativity inside black holes is still not fully understood, but many physicists believe that understanding black holes could provide clues about the nature of quantum gravity and the early universe.

In the 1970s, physicist Stephen Hawking made a groundbreaking discovery that linked quantum mechanics with black holes: the phenomenon now known as Hawking radiation. According to quantum mechanics, pairs of virtual particles constantly pop in and out of existence near the event horizon of a black hole. Occasionally, one of these particles falls into the black hole while the other escapes, causing the black hole to lose a tiny amount of mass. Over time, this process could lead to the black hole gradually evaporating. Hawking's discovery showed that black holes are not entirely black but emit radiation due to quantum effects, suggesting that quantum mechanics plays a crucial role in the behavior of black holes. This raises fascinating questions about the ultimate fate of black holes and whether information about the matter that falls into a black hole is preserved or lost.

Quantum cosmology also addresses the question of what might have existed before the Big Bang, or whether time itself even existed before the universe began. One possibility is that time, like space, is a feature that emerged from the quantum vacuum along with the universe. In this view, the concept of "before the Big Bang" may not have any physical meaning because time itself did not exist before the universe came into being. This idea is difficult to grasp because it challenges our everyday understanding of time as something that flows continuously, but it is a natural consequence of trying to apply quantum mechanics to the entire universe.

If quantum cosmology is correct, the universe we live in could be just one of many possible universes, each shaped by quantum effects at its origin. It also suggests that the universe is fundamentally governed by the same quantum principles that control the behavior of particles at the smallest scales, offering a tantalizing hint that quantum mechanics may hold the key to understanding the entire cosmos.

The Many-Worlds Interpretation

The Many-Worlds Interpretation (MWI) of quantum mechanics is one of the most fascinating and controversial interpretations of quantum theory. Proposed by American physicist Hugh Everett III in 1957, the Many-Worlds Interpretation offers a radical solution to the mysteries of quantum mechanics, particularly the problem of wave function collapse and the nature of measurement. According to this interpretation, whenever a quantum system is measured, the universe does not simply collapse to a single outcome, as in the Copenhagen Interpretation. Instead, all possible outcomes of the measurement actually occur—but each in its own separate, branching universe. This means that the universe constantly splits into multiple, parallel versions, with every possible outcome of a quantum event realized in a different universe.

To understand how the Many-Worlds Interpretation works, it is helpful to start with the concept of the wave function, which describes the state of a quantum system. In quantum mechanics, before a measurement is made, the wave function of a particle represents a superposition of all the possible states the particle can occupy. For example, an electron in an atom might be in a superposition of several energy levels, or a photon might be in a superposition of traveling through two slits in the famous double-slit experiment. In the Copenhagen Interpretation, the act of measurement causes the wave function to "collapse" into a single state, and the system assumes one definite outcome. Before this collapse, the different possible outcomes are merely probabilities.

In contrast, the Many-Worlds Interpretation rejects the idea of wave function collapse entirely. Instead, it suggests that the wave function never collapses; rather, it continues to evolve according to the equations of quantum mechanics. When a measurement is made, all possible outcomes of the measurement are realized, but each outcome happens in a separate, non-interacting universe. In this view, the entire universe exists in a massive superposition of states, and each time a quantum event occurs, the universe "splits" into multiple, parallel versions. Each version corresponds to one possible outcome of the quantum event.

For example, imagine an experiment where an electron can be in one of two possible states: spin-up or spin-down. Before the measurement is made, the electron is in a superposition of both spin-up and spin-down. According to the Many-Worlds Interpretation, when a measurement is made, the universe splits into two: in one universe, the electron is measured as spin-up, and in the other universe, it is measured as spin-down. Both outcomes occur, but they are realized in different branches of the universe. Each version of the universe contains a different outcome of the measurement, and the observer in each universe only experiences the outcome that exists in their particular branch.

The Many-Worlds Interpretation has several advantages over the Copenhagen Interpretation. One of the main advantages is that it avoids the problem of wave function collapse. In the Copenhagen Interpretation, wave function collapse is a somewhat mysterious process that occurs when a measurement is made, but the exact nature of this collapse is not well defined. By eliminating the need for collapse, the Many-Worlds Interpretation provides a more straightforward and deterministic explanation of quantum phenomena. The wave function evolves continuously, without any special process for measurement, and all possible outcomes are treated as equally real.

Additionally, the Many-Worlds Interpretation resolves the measurement problem in quantum mechanics. In the Copenhagen view, the act of measurement plays a special role in determining the outcome of a quantum event, but it is unclear what constitutes a "measurement" or why it has such a unique effect. The Many-Worlds Interpretation sidesteps this issue by suggesting that measurements do not cause any fundamental change to the system. Instead, measurement simply reveals the observer's place in one of the many branches of the universe. The observer is just one part of the overall quantum system, and their experience of a particular outcome is determined by the branch of the universe in which they find themselves.

The implications of the Many-Worlds Interpretation are profound. It suggests that there is not one single, unique universe but rather a vast multiverse made up of an enormous number of parallel universes. Every time a quantum event occurs, the universe splits, creating new branches that contain every possible outcome. This means that there could be an infinite number of parallel versions of reality, with each version corresponding to a different history or future. In one universe, you may have chosen to go left at a crossroads, while in another, you went right. Every decision, every random event, and every quantum fluctuation leads to a new branch of the universe.

Back to Schrodinger's cat. According to the Copenhagen Interpretation, before the box is opened, the cat is in a superposition of being both alive and dead. The act of observing the cat collapses the wave function, and the cat is either alive or dead when the box is opened. In the Many-Worlds Interpretation, however, both outcomes occur, but in different universes. In one universe, the observer finds the cat alive, and in another, the observer finds the cat dead. Both realities exist, but in separate, non-communicating branches of the multiverse.

The Role of Observation

In the quantum realm, the role of observation—specifically, the act of measurement—has always been a central and contentious topic. The concept that observation influences the outcome of a quantum system is a hallmark of the Copenhagen Interpretation of quantum mechanics. However, when we shift our focus to the Many-Worlds Interpretation (MWI) and the idea of a multiverse, the role of observation takes on a very different, though equally profound, character. Instead of causing the wave function to collapse into a single outcome, as the Copenhagen view suggests, the act of measurement in the multiverse becomes a branching point, where all possible outcomes are realized across different, non-interacting realities.

In the Many-Worlds Interpretation, the universe is constantly branching as quantum events unfold. Each possible outcome of a quantum event creates a new "branch" of the universe, and these branches correspond to different realities, or parallel universes, within the multiverse. What makes this interpretation so radical is that it posits that all possible outcomes occur, but each occurs in its own separate universe. Thus, observation does not "choose" a particular outcome, as it would in the Copenhagen Interpretation. Instead, the act of measurement reveals the observer's place in one of the many branches, without collapsing the wave function or eliminating the other possibilities.

One of the key philosophical questions that arises from this interpretation is the nature of the observer's experience. If every possible outcome occurs in a different branch of the multiverse, what does it mean for the observer's perception of reality? When we make a measurement, we only ever experience one outcome. For instance, if you flip a quantum coin and measure it, you might observe it as heads. In the Many-Worlds Interpretation, another version of you in a different branch of the multiverse sees tails. Both outcomes exist, but each version of you is unaware of the other. This raises deep questions about identity and what it means to be an observer in a branching multiverse.

From the perspective of an observer, the multiverse is fundamentally deterministic, even though individual outcomes appear probabilistic. Before making a measurement, you might think there is a 50% chance of observing heads and a 50% chance of observing tails. However, in the Many-Worlds Interpretation, the reality is that both outcomes will occur, but in different branches of the universe. To the observer in any given branch, the result seems random, because they are only aware of their own branch and cannot interact with or observe the outcomes in other branches. This is how the Many-Worlds Interpretation reconciles the apparent randomness of quantum mechanics with a deterministic, branching multiverse.

Implications of this view is that every decision and every quantum event leads to the creation of new universes. Imagine making a choice between two options—say, turning left or right at a crossroads. In the Many-Worlds Interpretation, both choices are made, but each one occurs in a different universe. In one branch, you turn left, and in another, you turn right. The act of observation or decision-making doesn't determine a single outcome but instead causes the universe to branch, creating a new reality for each possibility.

This idea has far-reaching consequences, not just for quantum events but for how we think about time, choice, and the structure of reality. It suggests that every time we make a measurement or decision, we are participating in the branching of the multiverse. In one sense, this could be seen as empowering: all possible versions of events do happen, and every possible version of "you" exists in some branch of the multiverse. Are the different versions of "you" in each branch truly separate individuals, or are they all part of a larger, interconnected whole? And what does this mean for our sense of free will, if all possible choices are realized in some version of the multiverse?

The role of observation in the multiverse also has implications for the concept of reality itself. In the classical view of physics, reality is objective and exists independently of observation. However, in the Many-Worlds Interpretation, reality becomes more fluid and dependent on the observer's position within the multiverse. Each observer experiences only one branch of the multiverse, but all branches are equally real. This challenges our classical understanding of a single, objective reality and suggests that reality is much more complex and layered than we typically assume.

Criticism and Debate

The multiverse theory, the idea that our universe is just one of potentially infinite parallel universes, has generated significant interest in both scientific and popular culture. While the concept offers a solution to some of the most perplexing questions in physics and cosmology, it also faces substantial criticism and debate. Both scientific and philosophical objections have been raised about the theory, with critics questioning its testability, its implications for the nature of science, and its philosophical coherence. These objections highlight the challenges of making a radical idea like the multiverse fit within the framework of empirical science.

One of the central criticisms of the multiverse theory is that it is inherently difficult, if not impossible, to test experimentally. In most formulations of the multiverse, whether arising from string theory, the inflationary model of cosmology, or the Many-Worlds Interpretation of quantum mechanics, the different universes are thought to be causally disconnected from each other. This means that no information can pass between them, and there is no direct way to observe or interact with other universes. From a scientific standpoint, this presents a major problem. If the multiverse cannot be observed, measured, or experimentally confirmed, how can it be considered a legitimate scientific theory? The lack of empirical evidence has led many critics to argue that the multiverse is not a scientifically meaningful concept but rather a philosophical or speculative idea.

This issue touches on a broader debate about the nature of scientific theories. According to philosopher of science Karl Popper, a theory must be falsifiable to be considered scientific. In other words, it must make predictions that can, in principle, be proven wrong through observation or experiment. Because the multiverse does not seem to offer any direct, testable predictions that could be falsified, many argue that it fails this criterion. As a result, some physicists and philosophers view the multiverse as speculative metaphysics rather than science. They contend that while the multiverse may be an intriguing concept, it does not belong in the domain of empirical science, which is based on testability and observational evidence.

Proponents of the multiverse theory counter this criticism by pointing out that some versions of the theory could have indirect observational consequences. For example, in the context of cosmic inflation, the multiverse might leave subtle imprints on the cosmic microwave background radiation, which could be detectable with highly sensitive instruments. In this view, while we may never directly observe another universe, the multiverse could still have observable effects on our own universe, allowing for some level of empirical testing. However, these potential observations remain speculative, and there is currently no conclusive evidence that directly supports the existence of a multiverse.

Another scientific objection to the multiverse theory concerns the principle of parsimony, or Occam's razor. This principle states that when faced with competing explanations for a phenomenon, we should prefer the simplest one that requires the fewest assumptions. Critics argue that the multiverse, by postulating an infinite or near-infinite number of universes, violates this principle by introducing a highly complex and unobservable structure to explain phenomena that could potentially be explained by simpler, more conventional theories. For instance, some argue that instead of invoking the existence of countless other universes, we should focus on refining our understanding of the physics within our own universe to explain the observed features of reality

Philosophically, the multiverse theory raises questions about the nature of reality and the concept of probability. One of the appealing aspects of the multiverse is that it offers a solution to the fine-tuning problem—the observation that the physical constants of our universe seem to be precisely tuned to allow for the existence of life.

In a multiverse, with different universes having different physical constants, the fact that we live in a universe capable of supporting life is not surprising because, in an infinite multiverse, there would inevitably be some universes with the right conditions for life to emerge. We just happen to live in one of them.

However, critics argue that this explanation is unsatisfying because it relies on a kind of "anthropic reasoning." The anthropic principle states that we observe the universe to be the way it is because we exist to observe it. While this may explain why we find ourselves in a life-friendly universe, it does not necessarily provide a deeper understanding of why the physical constants are the way they are. Some philosophers argue that invoking the multiverse to solve the fine-tuning problem is akin to "giving up" on finding a fundamental explanation for the values of these constants within our own universe. Instead of seeking a deeper understanding of why our universe has the properties it does, the multiverse offers a kind of statistical explanation that some find unsatisfactory.

The multiverse theory also raises concerns about the "measure problem." If there are an infinite number of universes, how can we meaningfully discuss probabilities or make predictions? In an infinite multiverse, it becomes difficult to define what is "typical" or "probable" because every possible outcome occurs an infinite number of times. This makes it challenging to make sense of statistical reasoning, which is crucial to much of physics. For example, if every possible outcome of a quantum event occurs in some branch of the multiverse, what does it mean to say that an event is "likely" or "unlikely"? This issue complicates the task of using the multiverse to make predictions about our own universe, and it has led some critics to argue that the multiverse is not a coherent or useful framework for understanding probability.

Quantum Field Theory

It is the theory that combines two foundational concepts in physics: quantum mechanics, which describes the behavior of particles at the smallest scales, and special relativity, which explains how objects move at high speeds, approaching the speed of light. QFT is essential for understanding the behavior of fundamental particles and their interactions, and it forms the basis of the Standard Model of particle physics, the theory that describes three of the four known fundamental forces in nature: electromagnetism, the strong nuclear force, and the weak nuclear force.

Quantum Field Theory (QFT) is one of the most important and successful frameworks in modern physics. It is the theory that combines two foundational concepts in physics: quantum mechanics, which describes the behavior of particles at the smallest scales, and special relativity, which explains how objects move at high speeds, approaching the speed of light. QFT is essential for understanding the behavior of fundamental particles and their interactions, and it forms the basis of the Standard Model of particle physics, the theory that describes three of the four known fundamental forces in nature: electromagnetism, the strong nuclear force, and the weak nuclear force.

At its core, quantum mechanics describes how particles behave at the atomic and subatomic scales. However, early quantum mechanics, developed in the 1920s and 1930s, was focused on particles like electrons and photons without fully taking into account the principles of special relativity. Special relativity, formulated by Albert Einstein in 1905, tells us that nothing can travel faster than the speed of light and that time and space are interconnected in a single framework known as spacetime. This posed a problem for quantum mechanics, which needed to account for the relativistic behavior of particles traveling at or near the speed of light, particularly those like photons, which always move at light speed.

Quantum Field Theory was developed to solve this problem by merging quantum mechanics with special relativity. The key idea of QFT is that instead of treating particles as individual, isolated objects, it treats them as excitations or disturbances in underlying fields. In this framework, the fundamental entities of nature are not particles themselves, but fields that exist throughout space and time. These fields are continuous and can be thought of as the medium through which particles are created and interact. For example, an electron is viewed not as a discrete point particle, but as a localized excitation in the electron field, and a photon is an excitation in the electromagnetic field.

The concept of fields in physics is not new—electromagnetic fields have been well understood since the 19th century. However, in QFT, all particles, not just photons, are described in terms of fields. Each type of particle has its own associated field: electrons have an electron field, quarks have quark fields, and so on. These fields are governed by the principles of quantum mechanics, which means that their excitations are quantized, giving rise to the discrete particles we observe.

One of the most important consequences of quantum field theory is that it allows for the creation and annihilation of particles. In classical mechanics, the number of particles is fixed, and they follow continuous trajectories through space. But in QFT, particles can be created or destroyed through interactions with their respective fields. For instance, in high-energy collisions, new particles can emerge as a result of the energy being converted into mass, in accordance with Einstein's famous equation $E=mc2$. This process is fundamental to our understanding of particle physics and plays a key role in phenomena such as particle-antiparticle pair creation and annihilation.

Another crucial aspect of QFT is the way it handles interactions between particles. In classical field theory, forces like electromagnetism are described by continuous fields, with charged particles interacting through the exchange of force carriers (like the electromagnetic field for electrons). In QFT, interactions between particles are described in a similar way, but the force carriers themselves are quantized. For example, in quantum electrodynamics (QED), the quantum field theory of electromagnetism, the electromagnetic force between charged particles is mediated by the exchange of virtual photons. These photons are not real, observable particles but are virtual particles that exist only for a fleeting moment during the interaction. The exchange of these virtual particles allows particles like electrons to exert forces on each other without ever coming into direct contact.

Quantum Field Theory goes beyond quantum electrodynamics to include other forces as well. For example, quantum chromodynamics (QCD) is the quantum field theory that describes the strong nuclear force, which binds quarks together inside protons and neutrons. In QCD, the strong force is mediated by particles called gluons, which, like photons in QED, are the force carriers for the interactions between quarks. The weak nuclear force, responsible for certain types of radioactive decay, is described by the electroweak theory, which is another quantum field theory that unifies the weak force and electromagnetism at high energies.

One of the remarkable achievements of Quantum Field Theory is its ability to make extremely precise predictions that match experimental results with astounding accuracy. For example, quantum electrodynamics (QED) has been tested to an extraordinary degree of precision, with theoretical predictions for the magnetic moment of the electron matching experimental measurements to many decimal places. This makes QFT one of the most successful scientific theories ever developed.

However, despite its success, QFT is not without its challenges. One of the most significant problems is that Quantum Field Theory struggles to incorporate gravity. While QFT provides an excellent description of three of the four fundamental forces—electromagnetism, the strong force, and the weak force—it does not yet provide a satisfactory theory of gravity. General relativity, which describes gravity as the curvature of spacetime, is fundamentally different from the quantum field theories that describe the other forces, and so far, physicists have been unable to merge these two frameworks into a single, unified theory of quantum gravity.

Gauge Theories and Symmetry

To explain gauge theories, we must first understand the concept of symmetry, a fundamental principle in physics that has guided the development of many theories.

In everyday life, symmetry is something we associate with balance or regularity—like the symmetry of a face or the regular pattern of a snowflake. In physics, symmetries are more abstract but equally important. They refer to transformations that leave the underlying physical laws unchanged. For example, the laws of physics do not change if you rotate an object or move it to a different location in space. This invariance under rotation or translation is a symmetry of the physical system.

Gauge theories are based on a particular kind of symmetry known as local symmetry. Unlike global symmetries, where transformations are uniform throughout space, local symmetries allow different transformations at each point in space and time. These transformations are governed by a set of rules called "gauge transformations." The central idea behind gauge theories is that these local symmetries correspond to the fundamental forces of nature.

To see how this works, consider electromagnetism, one of the most well-understood forces in nature. In the framework of quantum field theory, electromagnetism is described by a gauge theory called quantum electrodynamics (QED). The key insight of QED is that the electromagnetic force can be viewed as arising from the requirement that certain types of quantum fields, like the field associated with the electron, must respect a local symmetry. This symmetry is called a "gauge symmetry," and the associated gauge transformation involves a change in the phase of the electron's wavefunction at every point in space and time.

However, enforcing this local symmetry has profound consequences. It requires the existence of a new field that compensates for changes in the electron's phase. This field is nothing other than the electromagnetic field, and its quanta—photons—are the particles that mediate the electromagnetic force. Thus, the gauge symmetry, an abstract mathematical requirement, leads to the emergence of a real, observable force in nature.

This concept extends beyond electromagnetism. In fact, all of the known fundamental forces (except gravity) are described by gauge theories. The weak force, responsible for processes like radioactive decay, and the strong force, which binds quarks together inside protons and neutrons, are also governed by gauge theories. In the case of the weak and strong forces, the gauge symmetries are more complex, involving transformations that mix different types of particles together. These symmetries lead to the existence of other force-carrying particles, such as the W and Z bosons for the weak force and gluons for the strong force.

Gauge theories reveal a deep and elegant structure underlying the forces of nature. They show that what we perceive as fundamental forces are in fact manifestations of symmetries at the quantum level. The symmetry principles that govern the behavior of quantum fields not only predict the existence of force-carrying particles but also determine how those particles interact with matter.

In quantum field theory, symmetries do more than provide aesthetic appeal; they dictate the dynamics of particles and fields. The requirement that a theory be invariant under a particular gauge symmetry leads directly to the mathematical form of the interactions between particles. For example, the fact that the electromagnetic interaction is long-range and mediated by photons is a consequence of the specific gauge symmetry that governs electromagnetism.

Gauge theories also have profound implications for the unification of forces. In the Standard Model of particle physics, the electromagnetic, weak, and strong forces are all described by gauge theories, and physicists believe that these forces may have been unified into a single force in the early universe. This idea of unification suggests that the differences between the forces we observe today arose as the universe cooled and the underlying symmetries were broken.

Renormalization: Dealing with Infinities in Quantum Calculations

In the realm of quantum physics, calculations involving interactions between particles often give rise to a perplexing issue: infinities. These infinities occur when we try to compute certain physical quantities, such as the energy or charge of particles, using the rules of quantum field theory. At first glance, these infinities seem like a serious flaw in the theory, suggesting that quantum mechanics might be fundamentally broken. However, a technique called renormalization has been developed to deal with these infinities, allowing us to extract meaningful, finite results from otherwise divergent calculations.

To understand the need for renormalization, we need to consider how quantum field theory works. In this framework, particles interact by exchanging force-carrying particles, such as photons in the case of electromagnetism. These interactions are often described in terms of Feynman diagrams, which represent the various ways that particles can scatter off one another. Some of these diagrams correspond to relatively simple processes, while others involve more complex, higher-order processes where particles interact multiple times or even create and annihilate additional particles along the way.

When we calculate the probability of these interactions, we sum up contributions from all possible Feynman diagrams, including those representing higher-order interactions. But here's the problem: when we try to account for all of these possibilities, particularly those involving very high-energy interactions or particles interacting over very short distances, the result often becomes infinite. This clearly doesn't match with what we observe in experiments—after all, we don't see particles with infinite energy or infinite forces between them.

In the early days of quantum field theory, these infinities posed a serious challenge. How could a theory that produces nonsensical, infinite answers be reconciled with the accurate, finite predictions required by experiment? This is where renormalization comes in. It is a mathematical procedure that allows us to systematically eliminate these troublesome infinities and replace them with finite quantities that correspond to observable physical properties, such as the mass and charge of a particle.

The key idea behind renormalization is that the infinities we encounter in quantum field theory are not necessarily a problem with the theory itself, but rather a reflection of how we choose to describe the system. In particular, the raw, unrenormalized quantities that appear in our calculations—such as the bare mass or bare charge of a particle—are not the physical quantities we actually measure. Instead, these bare quantities include contributions from all possible interactions between the particle and the quantum field surrounding it.

For example, the charge of an electron is not simply the charge of the "bare" electron itself. It also includes the effects of virtual particles, such as photons, that surround the electron and momentarily "dress" it. These virtual particles can alter the observed charge and mass of the electron, depending on how closely we look. As we examine the system at higher energies or shorter distances, more virtual particles contribute, and the observed values can change.

Renormalization involves redefining the bare parameters in our theory in terms of the physical quantities that we can actually observe, such as the electron's measured charge and mass. In practical terms, this means we systematically subtract out the infinite contributions to the bare quantities, leaving behind only the finite, observable values. These observable values remain finite, even though the underlying theory might produce infinities at certain steps in the calculation.

One of the remarkable successes of renormalization is its application to quantum electrodynamics (QED), the quantum theory of the electromagnetic force. Early in the development of QED, physicists like Richard Feynman and Julian Schwinger used renormalization to show that, despite the presence of infinities in intermediate calculations, the theory could still make extremely accurate predictions about the behavior of electrons and photons. In fact, QED remains one of the most precisely tested theories in all of physics, with predictions matching experimental results to an extraordinary degree of accuracy.

The need for renormalization suggests that our description of particles and forces is not complete at all scales. At very high energies, or equivalently at very short distances, new physical phenomena might emerge that we don't yet fully understand. For example, many physicists believe that renormalization hints at the existence of new physics beyond the Standard Model, the current best framework for describing fundamental particles and forces.

In particular, the process of renormalization has led to the development of the renormalization group, a powerful mathematical tool that allows physicists to study how physical systems behave as we change the energy scale of our observations. This approach has had far-reaching implications, not only in particle physics but also in fields like condensed matter physics, where it has been used to understand phase transitions and other complex phenomena.

Pilot-Wave Theory

Pilot-wave theory, also known as the de Broglie-Bohm interpretation of quantum mechanics, offers a unique alternative to the standard interpretation of quantum mechanics, particularly the Copenhagen Interpretation. In the Copenhagen view, particles do not have definite properties such as position or momentum until they are measured, and the wavefunction, which describes the quantum system, collapses during measurement to produce a specific outcome. Pilot-wave theory, however, takes a very different approach. It postulates that particles always have definite positions and velocities, and their motion is guided by a "pilot wave" that determines how they evolve over time. In this interpretation, there is no wavefunction collapse, and the uncertainty we associate with quantum mechanics arises not from the fundamental nature of particles, but from our lack of knowledge about their exact initial conditions.

The origins of pilot-wave theory date back to the 1920s, when French physicist Louis de Broglie introduced the idea of matter waves. De Broglie proposed that particles, such as electrons, exhibit wave-like behavior, a hypothesis that was soon confirmed experimentally in the famous electron diffraction experiments. According to de Broglie, particles are associated with a guiding wave, or pilot wave, that governs their motion. This idea was initially well received, but it was later overshadowed by the rise of the Copenhagen Interpretation, which became the dominant framework for understanding quantum mechanics.

However, the pilot-wave concept was revived and expanded by physicist David Bohm in the 1950s, giving rise to what is now called the de Broglie-Bohm interpretation or simply Bohmian mechanics. Bohm's version of the theory provided a clearer mathematical foundation for de Broglie's original ideas and showed that the pilot-wave theory could reproduce all the same experimental predictions as standard quantum mechanics. In fact, pilot-wave theory is empirically indistinguishable from the Copenhagen Interpretation, meaning that both approaches give the same predictions for the outcomes of experiments. What distinguishes them is their underlying conceptual frameworks.

In pilot-wave theory, particles are real objects with well-defined trajectories, meaning they have specific positions and velocities at all times. These particles are guided by a wave, described by the quantum wavefunction, which influences their motion but does not collapse upon measurement. Instead of collapsing into a single state when observed, the wavefunction continues to exist and evolve, guiding the motion of all particles in the system. This is a significant departure from the standard interpretation, where the wavefunction represents the probabilities of different outcomes and collapses into one outcome when a measurement is made.

In pilot-wave theory, the wavefunction serves as a kind of "field" that permeates all of space, directing the motion of particles in a deterministic way. This means that, if we knew the initial conditions of the system—such as the exact positions and velocities of all the particles—we could, in principle, predict their future behavior with certainty. This contrasts with the standard view of quantum mechanics, which is inherently probabilistic, meaning that we can only calculate the likelihood of different outcomes, not determine them exactly. In this sense, pilot-wave theory restores a form of determinism to quantum mechanics, which is appealing to many physicists who are uncomfortable with the inherent randomness of the Copenhagen Interpretation.

A central feature of pilot-wave theory is that it explains quantum phenomena like interference and diffraction, not as the result of particles being in multiple states at once, but as the result of the interaction between the particles and the pilot wave. For example, in the famous double-slit experiment, a particle, such as an electron, passes through one of two slits, but its motion is influenced by the pilot wave, which passes through both slits and interferes with itself. This interference pattern in the pilot wave then guides the particle to specific regions on the detection screen, creating the characteristic interference pattern observed in the experiment. In this way, pilot-wave theory can explain the wave-like behavior of particles without resorting to the idea of superposition or wavefunction collapse.

In standard quantum mechanics, the concept of wavefunction collapse is somewhat mysterious, and the idea that particles exist in superpositions of states until measured can be difficult to reconcile with our everyday experience of reality. Pilot-wave theory, by contrast, offers a more classical view of the world, where particles have definite properties at all times, and their motion is governed by a guiding wave. This can make the quantum world seem less strange, as it removes the need for particles to "exist" in multiple states simultaneously.

However, pilot-wave theory is not without its challenges and criticisms. One of the main criticisms is that it introduces non-locality, meaning that particles can be influenced by distant events instantaneously, violating the principle that nothing can travel faster than the speed of light. In pilot-wave theory, the pilot wave exists throughout space and can instantaneously affect the motion of particles, even if they are far apart. This non-locality was famously demonstrated by John Bell in the 1960s, who showed that any theory that reproduces the predictions of quantum mechanics must be non-local. While non-locality is a feature of standard quantum mechanics as well (as seen in phenomena like quantum entanglement), it is more explicitly built into the framework of pilot-wave theory, and some physicists are uncomfortable with this aspect.

Unsolved Mysteries in Quantum Physics

Quantum Gravity

Quantum gravity represents one of the greatest challenges in modern theoretical physics. It is the quest to develop a single, unified theory that can reconcile the principles of quantum mechanics with those of general relativity. Both quantum mechanics and general relativity are incredibly successful in their respective domains: quantum mechanics governs the behavior of particles at the smallest scales, while general relativity describes gravity and the structure of spacetime on cosmic scales. However, when these two theories are applied together—such as in extreme environments like black holes or the early universe—they are found to be incompatible. The search for a theory of quantum gravity aims to bridge this divide and create a unified framework that can describe the universe on all scales.

At the heart of this challenge is the fundamentally different way that quantum mechanics and general relativity describe reality. Quantum mechanics is based on the concept of wave functions and probabilities, where particles exist in superpositions of states and exhibit behavior governed by uncertainty. In this realm, forces like electromagnetism are mediated by the exchange of particles called force carriers, such as photons. This approach works well for describing three of the four fundamental forces of nature: electromagnetism, the strong nuclear force, and the weak nuclear force. These forces are all described by quantum field theories, which treat particles and forces as manifestations of underlying fields that obey the rules of quantum mechanics.

General relativity, however, takes a very different approach to describing the fourth force: gravity. Instead of treating gravity as a force mediated by particles, general relativity views gravity as the result of the curvature of spacetime. Massive objects like stars and planets warp the fabric of spacetime around them, and this curvature affects the motion of other objects, which follow curved paths in response. This geometric description of gravity is both elegant and powerful, and it has been confirmed by countless experiments and observations, from the bending of starlight around the sun to the detection of gravitational waves from colliding black holes.

The problem arises when we try to apply both quantum mechanics and general relativity to situations where both gravity and quantum effects are important—such as near black holes or at the moment of the Big Bang. In these extreme environments, the gravitational field is incredibly strong, and spacetime itself becomes highly curved. At the same time, quantum mechanics predicts that particles and fields exhibit fluctuations, meaning that spacetime cannot remain smooth and continuous at these scales. Instead, it may become subject to quantum fluctuations, leading to a "foamy" or "grainy" structure at the smallest scales. Unfortunately, when physicists attempt to combine quantum mechanics and general relativity in these regimes, the mathematical frameworks of the two theories break down, resulting in nonsensical infinities.

One of the primary goals of quantum gravity is to find a way to describe gravity within the framework of quantum mechanics. In other words, physicists are searching for a theory in which gravity is quantized, just like the other fundamental forces. If successful, this theory would unite the four forces of nature under a single, coherent framework—a goal sometimes referred to as the "Theory of Everything." A quantum theory of gravity would also offer new insights into the behavior of spacetime at the smallest scales, possibly shedding light on the true nature of space, time, and matter.

There have been several promising approaches to the problem of quantum gravity, though none have yet been proven or widely accepted. One of the most well-known is string theory, which proposes that the fundamental building blocks of the universe are not point-like particles, but tiny, vibrating strings. According to string theory, different types of particles arise from different vibrational modes of these strings. Crucially, string theory naturally incorporates gravity into its framework by predicting the existence of a particle called the graviton, which mediates the gravitational force in the same way that photons mediate the electromagnetic force. In string theory, gravity is no longer a purely geometric phenomenon but is described by the exchange of gravitons, just as quantum mechanics predicts for other forces.

String theory is appealing because it offers a way to unite quantum mechanics and gravity within a single framework.

However, it also faces significant challenges. For one, string theory requires the existence of extra spatial dimensions beyond the familiar three, and these extra dimensions are typically assumed to be "curled up" so small that they are unobservable at human scales. This makes string theory difficult to test experimentally, as it predicts phenomena that are far beyond the reach of current experiments. Additionally, string theory is highly complex and has yet to make any definitive predictions that could be tested in the laboratory, which has led some critics to argue that it remains speculative.

Another approach to quantum gravity is loop quantum gravity, which takes a very different path from string theory. Instead of introducing extra dimensions or new particles, loop quantum gravity attempts to directly quantize spacetime itself. According to loop quantum gravity, space is not a smooth, continuous fabric but is made up of tiny, discrete "loops" or "chunks" at the Planck scale, which is many orders of magnitude smaller than an atom. These loops form a kind of network that gives rise to the structure of spacetime. Loop quantum gravity also predicts that time is quantized, meaning that it may consist of individual moments rather than flowing continuously. This approach preserves many of the key insights of general relativity while applying quantum principles to the structure of spacetime itself.

Loop quantum gravity has been successful in addressing some of the key problems of quantum gravity, particularly in describing what happens to spacetime near singularities, such as the centers of black holes or the beginning of the universe. In some models, loop quantum gravity suggests that the singularity at the center of a black hole is replaced by a finite, quantum-corrected region, possibly allowing information to escape from black holes—a solution to the so-called "black hole information paradox." Additionally, loop quantum gravity offers a potential resolution to the problem of the Big Bang singularity, suggesting that the universe may have undergone a "bounce" from a previous contracting phase.

Despite these advances, both string theory and loop quantum gravity remain incomplete. Neither theory has been fully developed to the point where it can be rigorously tested by experiments or fully explain all aspects of quantum gravity. Moreover, there are other approaches to quantum gravity that offer competing ideas, such as causal dynamical triangulations or asymptotic safety, each with its own advantages and challenges. The search for a theory of quantum gravity is still very much an active area of research, and physicists continue to explore new ideas in the hope of achieving a unified understanding of the universe.

Dark Matter and Quantum Mechanics

Dark matter is one of the greatest mysteries in modern cosmology and astrophysics. It is an invisible form of matter that makes up roughly 85% of the total mass of the universe, yet it cannot be seen directly because it does not interact with light. We only know of its existence through its gravitational effects on visible matter, such as galaxies and galaxy clusters. For decades, scientists have been trying to understand the nature of dark matter, and one of the possible avenues of exploration is its connection to quantum mechanics.

Quantum mechanics governs the behavior of matter at the smallest scales, from individual particles like electrons and quarks to composite structures such as atoms and molecules. Given that dark matter is a fundamental part of the universe, many researchers believe that its origins and properties may also be rooted in quantum mechanics. This raises the intriguing question: Could dark matter be composed of particles that obey the laws of quantum physics? And if so, what are the potential quantum properties of dark matter?

One of the leading candidates for dark matter is a class of hypothetical particles known as Weakly Interacting Massive Particles, or WIMPs. These particles would be massive enough to account for the gravitational effects we observe on cosmic scales but would interact with regular matter only through the weak nuclear force and gravity. Because WIMPs would interact so rarely with ordinary matter, they would be incredibly difficult to detect directly, but they would still follow the principles of quantum mechanics.

The search for WIMPs has focused on detecting the faint interactions they might have with ordinary matter in highly sensitive underground detectors. These detectors are designed to be shielded from cosmic rays and other sources of interference, so any signal they pick up might be due to dark matter particles. Although no definitive detection of WIMPs has yet been made, experiments continue to push the boundaries of sensitivity. If WIMPs are eventually found, they would represent a tangible connection between quantum mechanics and dark matter, revealing that dark matter is composed of a new type of quantum particle.

Another intriguing possibility is that dark matter could be made up of axions, another class of hypothetical particles that are much lighter than WIMPs. Axions were originally proposed as a solution to a problem in quantum chromodynamics (QCD), the theory that describes the strong nuclear force. However, they also have properties that make them suitable candidates for dark matter. Like WIMPs, axions would interact very weakly with ordinary matter, making them difficult to detect. If axions exist, they would also obey quantum mechanical laws and could exhibit wave-like properties due to their low mass. In fact, some theories suggest that axions could form a "dark matter condensate," where many axions behave collectively in a coherent quantum state, similar to how particles in a Bose-Einstein condensate behave at extremely low temperatures.

In addition to WIMPs and axions, there are other theoretical particles that might explain dark matter, and many of these ideas are rooted in quantum field theory, the framework that combines quantum mechanics with special relativity. For example, some theories propose that dark matter could be composed of sterile neutrinos, which are heavier versions of the familiar neutrinos that we know from the Standard Model of particle physics. Sterile neutrinos would interact only through gravity, making them excellent dark matter candidates. These particles, if they exist, would also follow the principles of quantum mechanics, adding another potential quantum layer to the dark matter puzzle.

One of the most exciting possibilities is that dark matter may not only obey quantum mechanics but also have uniquely quantum properties that distinguish it from ordinary matter. For instance, dark matter might exhibit quantum interference and entanglement—phenomena that have no classical analogs but are central to quantum physics. In some speculative theories, dark matter could even be linked to the multiverse or higher-dimensional spaces, ideas that arise from attempts to unify quantum mechanics with general relativity and string theory.

If dark matter does have quantum mechanical properties, this could help explain why it has been so difficult to detect. In quantum mechanics, particles can exist in a superposition of states, meaning they do not have definite properties until they are measured. It is possible that dark matter particles exhibit similar behavior, making them elusive to current detection methods. Alternatively, dark matter could interact with ordinary matter through quantum tunneling, a process where particles can pass through energy barriers that would be impassable in classical physics. If dark matter particles are quantum mechanical in nature, understanding their wave functions and interactions may require entirely new types of experiments and detectors.

Another potential connection between dark matter and quantum mechanics lies in the nature of dark energy, a mysterious force that is driving the accelerated expansion of the universe. Some theories suggest that dark energy and dark matter may be related, with both arising from quantum fluctuations in the vacuum of space. These fluctuations are predicted by quantum field theory, and while they are usually too small to have a noticeable effect on everyday scales, they could become significant in the context of the entire universe. If dark energy and dark matter share a common quantum origin, this could lead to a deeper understanding of both phenomena.

Finally, quantum mechanics might offer new ways of studying the large-scale distribution of dark matter in the universe. For example, quantum principles could be applied to the formation of dark matter halos around galaxies, which help to explain the rotation curves of galaxies and the distribution of visible matter. In this sense, quantum mechanics might not only explain the microscopic properties of dark matter particles but also their macroscopic effects on the structure of the universe.

The Arrow of Time

The "arrow of time" is a concept that refers to the fact that time seems to move in a single direction: from the past to the future. We experience the world as a sequence of events progressing forward, and while we can remember the past, we cannot reverse the process and recall the future. This directionality of time is one of the most fundamental aspects of our experience of the world, yet it remains one of the most profound mysteries in physics. Understanding why time moves forward has puzzled scientists for centuries, and quantum mechanics may offer new insights into this age-old question.

In classical physics, the laws of motion—whether from Newtonian mechanics or Einstein's theory of relativity—are time-symmetric. This means that they work just as well in reverse as they do going forward. If you were to watch a video of billiard balls colliding, for example, the physical laws that govern the motion of the balls would apply equally whether the video is played forwards or backwards. Nothing in these laws tells us why time should move in only one direction, and in this sense, classical physics does not provide an explanation for the arrow of time.

The key to understanding the arrow of time lies in the second law of thermodynamics, a principle from the field of thermodynamics that states that the total entropy of a closed system—essentially, a measure of disorder—always increases over time. Entropy is often associated with the degree of randomness or disorder in a system. For example, when you drop an ice cube into a glass of warm water, the ice melts, and the water and ice eventually reach the same temperature. The system has moved from a state of lower entropy (when the ice and water were at different temperatures) to a state of higher entropy (where everything is evenly distributed). Once the ice melts, the system doesn't spontaneously revert back to a state where the ice re-forms and the water cools down—time moves forward, and entropy increases.

The second law of thermodynamics gives us a thermodynamic arrow of time: the past is the state of lower entropy, and the future is the state of higher entropy. This law provides a statistical explanation for the direction of time, suggesting that the arrow of time is a consequence of the universe evolving from a highly ordered state (such as the Big Bang) to increasingly disordered states. But even with this understanding, there remains the question of why the universe started in such a low-entropy state and why time flows in only one direction.

This is where quantum mechanics might offer some answers. Quantum mechanics describes the behavior of particles at the smallest scales and introduces new ideas about the nature of time and measurement. In the quantum world, particles exist in superpositions of states, where they can be in multiple positions or have multiple energies at once, until they are measured. The process of measurement in quantum mechanics plays a key role in defining outcomes, and some physicists believe that the measurement process itself might be linked to the arrow of time.

One possible connection between quantum mechanics and the arrow of time is found in the concept of "decoherence." In quantum mechanics, a system can exist in a superposition of different states, but when it interacts with its environment—such as when it is measured or observed—these superpositions break down, and the system assumes a single, definite state. This process is known as decoherence, and it is often thought of as the quantum equivalent of "measurement." When decoherence occurs, the system loses its quantum coherence, and classical outcomes emerge. For example, an electron that might have been in multiple locations at once collapses into a single location when it is observed.

Decoherence introduces an asymmetry into quantum processes that may help explain the arrow of time. Before decoherence occurs, a quantum system can evolve in a time-symmetric way, meaning that its evolution could be reversed, just like in classical physics. But after decoherence, the system takes on a definite state, and the process is no longer time-symmetric—time now has a direction. Some physicists suggest that the arrow of time might emerge from this transition between quantum coherence (where time is reversible) and decoherence (where time flows in one direction).

Another way quantum mechanics might contribute to our understanding of the arrow of time is through the concept of quantum entanglement. When two particles become entangled, their states are correlated, even if they are separated by vast distances. If you measure one particle, the state of the other particle is immediately determined, no matter how far apart they are. Entanglement appears to defy the classical idea of locality and suggests that the underlying nature of time and space might be different at the quantum level. Some researchers believe that the way entanglement operates could provide clues to the flow of time, especially in highly entangled systems like the early universe.

Quantum cosmology, which combines quantum mechanics with cosmological models of the universe, also offers intriguing ideas about time. One possibility is that the arrow of time is a consequence of the quantum state of the universe at the moment of the Big Bang. The early universe may have existed in a quantum superposition of many possible states, and as it expanded and evolved, decoherence set in, creating the distinction between the past and the future. In this view, time's arrow could be a product of how the universe transitioned from its quantum origins to the classical world we experience today.

Moreover, the quest to unite quantum mechanics with general relativity in a theory of quantum gravity may further illuminate the nature of time. In general relativity, time is part of the fabric of spacetime, and it is shaped by the distribution of matter and energy. In quantum mechanics, time is treated as an external parameter, not something that is affected by the system being studied. A theory of quantum gravity would need to reconcile these two views, potentially offering a more fundamental understanding of time itself. Some theorists speculate that time, as we know it, might emerge from more basic quantum processes, just as temperature emerges from the motion of particles in thermodynamics.

Despite the progress made in understanding the arrow of time, it remains a deeply challenging problem. Quantum mechanics provides new tools and concepts that could help explain why time flows in one direction, but a complete explanation is still out of reach. The relationship between quantum processes like decoherence, entanglement, and the thermodynamic arrow of time remains an active area of research, and future discoveries in quantum physics and cosmology could offer more definitive answers.

Quantum Tunneling

Quantum tunneling describes the ability of particles to pass through energy barriers that they should not be able to cross according to the laws of classical physics. This process defies our everyday understanding of how particles behave, and it has puzzled scientists since it was first discovered. Even more intriguing is the fact that quantum tunneling appears to occur instantaneously.

In quantum tunneling, the particle doesn't "climb" over the barrier as it would in classical mechanics. Instead, it "tunnels" through the barrier, appearing on the other side without ever having enough energy to surmount the obstacle in the classical sense. The wave function allows for a small probability that the particle can be found beyond the barrier, and as a result, some particles will pass through, seemingly violating the classical laws of physics. This phenomenon is not just theoretical; it has been observed in numerous experiments and plays a key role in various physical processes, from the nuclear fusion reactions that power the sun to the operation of modern electronic devices like tunnel diodes and scanning tunneling microscopes.

One of the most puzzling aspects of quantum tunneling is the apparent instantaneous nature of the process. When particles tunnel through a barrier, experiments suggest that they do so without taking any measurable amount of time. In other words, the particle seems to vanish from one side of the barrier and reappear on the other side instantaneously, as if it bypasses the constraints of time altogether. This is deeply counterintuitive, as it challenges our classical understanding of how objects move through space and time.

The exact mechanism of how particles traverse barriers remains one of the open questions in quantum mechanics, but there are several important points to consider when thinking about quantum tunneling. First, the concept of time in quantum mechanics is not as straightforward as it is in classical physics. In classical mechanics, time flows continuously, and objects move through space in well-defined trajectories. But in quantum mechanics, particles do not follow definite paths; instead, they exist in a probabilistic haze, described by their wave functions. Because of this, it is difficult to define a precise "tunneling time" in the same way that we might measure how long it takes a classical particle to move from one point to another.

Moreover, the idea that quantum tunneling occurs "instantaneously" may stem from the fact that, in many experiments, the time it takes for a particle to tunnel through a barrier is so short that it cannot be measured by conventional means. This doesn't necessarily mean that the particle moves faster than the speed of light or violates any fundamental laws of physics. Instead, it points to the fact that the rules governing quantum behavior differ significantly from those that govern classical objects. While the particle's presence on both sides of the barrier might appear to be instantaneous, the underlying quantum processes are governed by probabilities, not deterministic paths or speeds.

Quantum tunneling is a deeply probabilistic phenomenon. The probability of a particle tunneling through a barrier depends on several factors, including the width and height of the barrier and the energy of the particle. Thicker or higher barriers make tunneling less likely, while thinner or lower barriers increase the probability. However, no matter how high the barrier, there is always a non-zero probability that the particle will tunnel through, which highlights the fundamental unpredictability and non-determinism inherent in quantum mechanics.

While quantum tunneling defies classical intuition, it plays a crucial role in many natural and technological processes. For example, in stars like our sun, nuclear fusion occurs because protons—positively charged particles that repel each other due to their electromagnetic force—can quantum tunnel through the repulsive barrier and come close enough to fuse together, releasing vast amounts of energy. Without quantum tunneling, the temperatures required for nuclear fusion in stars would be far higher than what we actually observe, making life as we know it impossible.

Quantum tunneling is also essential for modern electronics. In semiconductors, tunnel diodes exploit quantum tunneling to allow current to flow through thin barriers in a controlled way, enabling high-speed switching and signal processing in devices like computers and smartphones. Similarly, the scanning tunneling microscope (STM), a powerful tool for imaging surfaces at the atomic scale, relies on quantum tunneling to detect tiny currents that occur when electrons tunnel between the microscope's tip and the surface being studied.

Quantum Consciousness Theory

Quantum consciousness theory is an idea that attempts to explain consciousness—the subjective experience of being aware—using principles from quantum mechanics. It is a speculative and highly controversial area of research that lies at the intersection of physics, neuroscience, and philosophy. Proponents of quantum consciousness argue that classical physics alone cannot account for the full nature of consciousness and that quantum mechanics may hold the key to understanding how consciousness arises in the brain. It is important to note that it remains highly debated, and there is no conclusive scientific evidence to support the claim that quantum processes are directly responsible for consciousness.

To understand quantum consciousness theory, it is essential to first examine the challenges associated with explaining consciousness using classical neuroscience. The brain, with its billions of neurons and trillions of connections, is typically understood as a complex, classical system that operates according to the laws of classical physics. Neurons communicate with each other via electrical and chemical signals, forming intricate networks that give rise to thoughts, perceptions, emotions, and behaviors. However, many philosophers and scientists argue that there is a "hard problem" of consciousness—that is, explaining how subjective experience, or "qualia" (such as the redness of red or the pain of a headache), arises from the physical processes in the brain.

Classical physics, which operates on deterministic principles and well-defined laws, has so far struggled to explain the subjective, first-person nature of consciousness. This gap has led some researchers to suggest that the answer may lie in quantum mechanics, a theory that describes the behavior of particles at the smallest scales and includes phenomena such as superposition, entanglement, and wave function collapse. Quantum mechanics introduces indeterminism, probabilities, and strange connections between distant particles—features that seem very different from the deterministic and mechanistic approach of classical physics. Some proponents of quantum consciousness theory argue that the brain may utilize these quantum processes in ways that could help explain the elusive nature of consciousness.

One of the most well-known quantum consciousness theories was proposed by physicist Roger Penrose and anesthesiologist Stuart Hameroff. Their theory, known as Orchestrated Objective Reduction (Orch-OR), suggests that quantum processes in the brain's microtubules—a component of the cytoskeleton in neurons—are involved in generating consciousness. Microtubules are tiny, hollow tubes found within the cells of neurons, and Penrose and Hameroff hypothesize that quantum coherence could occur within these structures. According to their theory, quantum superpositions within microtubules are orchestrated by biological processes, and the collapse of these superpositions results in conscious experience.

The Orch-OR theory builds on Penrose's earlier work in quantum mechanics and his philosophical views on the limits of computation. Penrose argued that consciousness cannot be explained by traditional computation alone—such as the kind performed by computers or classical neurons—because human consciousness involves non-algorithmic thought processes that go beyond mechanical computation. He speculated that quantum mechanics, with its probabilistic nature and potential for non-local connections, might provide the framework needed to understand consciousness. Hameroff, a neuroscientist, then suggested that microtubules within neurons could be the biological structures capable of supporting quantum processes. Together, they proposed that the quantum states in microtubules are reduced or "collapsed" through a mechanism they called "objective reduction," which Penrose had independently theorized could occur at the quantum level.

While the Orch-OR theory is certainly imaginative, it has faced significant criticism from both physicists and neuroscientists. One of the main criticisms is that the brain is a warm, wet, and noisy environment—conditions that are typically unfavorable for sustaining quantum coherence, which is delicate and usually requires isolation from external disturbances. In most physical systems, quantum coherence is easily disrupted by interactions with the environment, a process known as decoherence. Critics argue that it is highly unlikely that quantum coherence could be maintained in the brain long enough to influence neural processes, particularly in microtubules, where thermal noise and biochemical activity would be expected to destroy any quantum effects quickly.

Moreover, while Penrose and Hameroff's theory attempts to connect quantum mechanics with consciousness, there is little experimental evidence to support the claim that quantum processes play a direct role in how the brain generates conscious experience. Most neuroscientists believe that consciousness can be understood within the framework of classical physics, based on known principles of neuroscience, such as neural networks, synaptic plasticity, and information processing in the brain. From this perspective, quantum mechanics is unnecessary for explaining how the brain works or how consciousness emerges.

Despite these criticisms, quantum consciousness theories continue to capture the imagination of some researchers and the public. One reason for this is the mysterious and counterintuitive nature of quantum mechanics itself. Quantum phenomena like superposition (where particles exist in multiple states at once) and entanglement (where particles are instantaneously connected across space) challenge our classical understanding of reality. These strange properties of the quantum world make it tempting to connect them to the mystery of consciousness, which is itself a difficult and poorly understood phenomenon.

Additionally, the fact that consciousness appears to involve subjective experience, awareness, and a sense of "being" that transcends the mechanistic workings of neurons has led some to seek explanations outside the boundaries of classical physics. For proponents of quantum consciousness, the unique and perplexing nature of quantum mechanics seems like a natural candidate for explaining the uniqueness of conscious experience.

Beyond Orch-OR, there are other speculative ideas about quantum consciousness. Some researchers have suggested that quantum entanglement could play a role in how different regions of the brain communicate with each other, allowing for the rapid integration of information necessary for conscious thought. Others have proposed that quantum processes might be involved in decision-making, creativity, or free will, although these ideas remain largely untested.

New Frontiers

Quantum Computing

Quantum computing is a revolutionary approach to computation that takes advantage of the principles of quantum mechanics, the branch of physics that governs the behavior of particles at the smallest scales. Unlike classical computers, which process information using bits—units that can be either a 0 or a 1—quantum computers use quantum bits, or qubits. The difference between these two types of bits is what makes quantum computers fundamentally different, and potentially much more powerful, than classical ones.

To understand the concept of a qubit, we need to first recall how classical bits work. In a classical computer, everything is ultimately represented as a sequence of 0s and 1s. These binary digits, or bits, are processed through electrical circuits, which perform operations to manipulate the bits according to the instructions of a program. While classical computers can perform billions of operations per second, they are limited by the fact that each bit can only exist in one of two states: 0 or 1, but never both at the same time.

Qubits, on the other hand, are governed by the principles of quantum mechanics, which allows them to exist in a superposition of states. A qubit is not restricted to being just 0 or 1; it can be in a state that represents both 0 and 1 simultaneously, in various combinations. This superposition of states is one of the key features that gives quantum computers their power. With just a few qubits, a quantum computer can process a vast number of possible outcomes at once, whereas a classical computer would need to process each possibility sequentially.

Another important concept in quantum computing is entanglement, a phenomenon where qubits become interconnected in such a way that the state of one qubit is directly related to the state of another, no matter how far apart they are. When qubits are entangled, changes to one qubit will instantaneously affect the other, even if they are separated by large distances. This property allows quantum computers to perform certain types of calculations in ways that are impossible for classical computers, because entangled qubits can work together in ways that classical bits cannot.

In addition to superposition and entanglement, quantum computers also rely on quantum interference to extract useful information from the multitude of possibilities they explore. In a quantum computation, different possible states of the qubits can interfere with each other, much like waves in water. This interference can be harnessed to amplify the correct answers to a problem while canceling out the incorrect ones. This selective interference is a delicate process, but it is what allows quantum computers to solve complex problems much more efficiently than classical computers.

While the theoretical principles behind quantum computing are well-established, building practical quantum computers is still a major technical challenge. Qubits are extremely fragile and can be easily disrupted by their environment, a problem known as quantum decoherence. This makes it difficult to maintain stable qubits for long enough to perform useful computations. Scientists and engineers are actively working on ways to address these challenges, and several different approaches to building quantum computers are being explored, including systems based on trapped ions, superconducting circuits, and even particles of light, called photons.

Despite these challenges, quantum computers hold the promise of transforming many fields of science and technology. For certain types of problems, such as factoring large numbers or simulating the behavior of complex molecules, quantum computers could potentially outperform classical computers by many orders of magnitude. This could have profound implications for cryptography, where the security of many encryption systems relies on the difficulty of factoring large numbers. It could also revolutionize fields like materials science and drug discovery, where simulating the behavior of molecules and atoms requires enormous computational power.

One of the most well-known quantum algorithms is Shor's algorithm, which can efficiently factor large numbers and has important implications for cryptography. Another is Grover's algorithm, which provides a way to search through unsorted data much faster than any classical algorithm. These algorithms demonstrate the potential of quantum computing, but they are just the beginning. Researchers are still discovering new algorithms that could take advantage of quantum principles in ways that we are only starting to understand.

It is important to recognize that quantum computers are not meant to replace classical computers for all tasks. For many everyday applications, classical computers are perfectly sufficient and, in many cases, more practical. However, for specific problems that involve large amounts of data, complex simulations, or cryptographic challenges, quantum computers could provide solutions that are out of reach for classical machines.

Quantum Biology

Quantum biology is an emerging field that seeks to explore and understand how quantum mechanical phenomena—typically associated with the microscopic world of atoms and particles—play a role in biological systems. At first glance, biology and quantum physics might seem to operate on very different scales. Biology deals with living organisms and the chemical processes that sustain them, while quantum mechanics deals with subatomic particles, where the rules of classical physics no longer apply. However, recent research suggests that quantum effects could influence some of the most fundamental processes in life, hinting at a deep connection between these seemingly unrelated domains.

The foundation of quantum biology lies in the idea that certain biological processes may rely on quantum effects like **superposition**, **entanglement**, and **tunneling**, which are typically observed in the quantum realm. These processes are often incredibly efficient and precise, leading researchers to investigate whether the unusual properties of quantum mechanics are at work within living organisms.

One of the most well-known examples where quantum biology is believed to play a role is in **photosynthesis**, the process by which plants and some bacteria convert sunlight into chemical energy. In this process, light energy is absorbed by molecules called chlorophyll, which then pass the energy through a series of proteins to a reaction center, where it is converted into a usable form of energy. What's remarkable about this energy transfer is its near-perfect efficiency. Researchers have found that the energy appears to travel through multiple pathways simultaneously, as if it were "testing" all possible routes to the reaction center before choosing the most efficient one. This behavior is reminiscent of quantum superposition, where particles can exist in multiple states or locations at the same time. The implication is that quantum effects might help optimize the process of energy transfer in photosynthesis, allowing plants to capture sunlight with incredible precision.

Another fascinating area of quantum biology is **enzyme catalysis**, the process by which enzymes accelerate chemical reactions in living organisms. Enzymes are critical to countless biological functions, from digesting food to replicating DNA. In some cases, researchers have suggested that quantum tunneling—a phenomenon where particles pass through energy barriers they shouldn't be able to overcome according to classical physics—could be responsible for the high efficiency of certain enzyme reactions. This quantum effect may allow particles, such as protons or electrons, to "tunnel" through energy barriers, speeding up the reaction in ways that classical mechanics can't explain.

Bird navigation offers another intriguing example of quantum biology. Some species of birds are able to migrate thousands of miles, relying on the Earth's magnetic field to guide their way. While it has long been known that birds have some kind of internal compass, the exact mechanism remained mysterious. Recent research suggests that birds may rely on quantum entanglement within proteins in their eyes, specifically a protein called cryptochrome. This protein may be sensitive to magnetic fields through a quantum process that affects the spin states of electrons. If this hypothesis is correct, it means that birds are able to "sense" the Earth's magnetic field due to quantum effects at the molecular level, enabling their remarkable navigational abilities.

The potential involvement of quantum mechanics in **olfaction**—the sense of smell—is another area of active investigation. Traditional theories of smell suggested that olfactory receptors in the nose detect molecules based on their shape, much like a lock and key. However, this theory alone has not been able to explain certain phenomena, such as why molecules with similar shapes can smell different or why molecules with different shapes can smell the same. Some researchers propose that quantum tunneling might be involved in the detection of smells. According to this theory, when a molecule binds to an olfactory receptor, an electron in the receptor might tunnel through a barrier if the molecule's vibrational frequency matches the energy difference required for the electron to jump. This vibrational theory of smell suggests that quantum processes could help determine the odors we perceive.

While quantum biology is still a relatively young field, it is increasingly attracting attention as scientists explore the possibility that life itself might exploit the strange rules of the quantum world. If quantum mechanics is indeed at play in these biological systems, it could represent a fundamentally new way of understanding biological processes—one that integrates physics, chemistry, and biology in a way that challenges traditional boundaries.

Despite its promise, quantum biology also faces significant challenges. One of the main questions is how quantum effects, which are typically very delicate and easily disrupted by environmental noise, can persist in the warm, wet, and chaotic environment of living cells. In laboratory experiments, quantum phenomena often require extreme conditions, such as very low temperatures and highly controlled environments, to be observed. However, biological systems operate at room temperature and are full of random fluctuations. How quantum effects can survive and even thrive under these conditions is still a subject of intense study.

Quantum biology may eventually provide answers to some of the most puzzling questions about life and its origins. For instance, researchers are exploring whether quantum effects played a role in the emergence of life on Earth, particularly in processes like the replication of DNA or the development of early metabolic pathways. These investigations could reveal new insights into how life first arose from the complex interplay of matter and energy.

Quantum Sensors and Metrology

Quantum sensors and quantum metrology represent cutting-edge fields that take advantage of the principles of quantum mechanics to vastly improve the precision and accuracy of measurements. While traditional measurement tools have advanced considerably over the years, quantum mechanics offers new ways of measuring time, space, and physical properties with unprecedented sensitivity. This is achieved by exploiting uniquely quantum phenomena like superposition, entanglement, and tunneling, which allow for finer control and manipulation of measurement systems at the atomic and subatomic levels.

At the heart of quantum sensing is the concept of **superposition**, where a quantum system, such as a particle or an atom, can exist in multiple states simultaneously. Classical systems, by contrast, are bound by the limits of binary choices—they can be in one state or another, but not both. Quantum systems, however, are much more flexible. In a quantum sensor, superposition can be used to create a range of possible outcomes that, when carefully measured, can detect even the smallest changes in the environment. This capability allows quantum sensors to be more sensitive to variations in physical quantities such as magnetic fields, temperature, and time.

An example of this can be found in **atomic clocks**, one of the most precise instruments ever developed. Traditional atomic clocks rely on measuring the oscillations of atoms to keep time, but quantum mechanics has enabled the development of next-generation clocks that are orders of magnitude more accurate. These quantum clocks use superposition and other quantum effects to reduce the uncertainties inherent in measurement, allowing them to measure time with almost unimaginable precision. To put it into perspective, a quantum atomic clock could lose only a second over the course of billions of years. This level of precision has profound implications for fields like navigation, communication, and even the synchronization of global financial systems.

Entanglement, another cornerstone of quantum mechanics, is also crucial to the power of quantum sensors. Entangled particles are deeply connected in such a way that the state of one particle instantaneously influences the state of another, regardless of the distance between them. In quantum sensing, entanglement allows information to be shared between particles in a way that can greatly enhance the precision of measurements. This is because entangled particles can essentially "work together," amplifying the signal being measured and reducing the noise that might otherwise obscure it. The result is a sensor that can detect extraordinarily subtle shifts in the environment, beyond what classical devices can achieve.

One area where entanglement has shown promise is in **quantum magnetometry**, the measurement of magnetic fields. Quantum magnetometers, which can detect tiny changes in magnetic fields using entangled particles, are already being used in applications ranging from medical imaging, such as MRI scans, to geological exploration, where detecting small magnetic anomalies can reveal valuable information about underground structures. Quantum magnetometers can also play a role in fundamental physics experiments, where detecting minute changes in magnetic fields can help test theories about the nature of the universe.

Quantum sensors also excel in the field of **gravitational wave detection**, where they are used to measure ripples in spacetime caused by distant astronomical events like the collision of black holes. Gravitational waves are extraordinarily weak signals, so detecting them requires instruments of extreme sensitivity. Quantum technologies have allowed for significant improvements in these detectors by reducing the noise that can obscure the faint signals of gravitational waves. The application of quantum techniques has already contributed to groundbreaking discoveries in astrophysics, with the first direct detection of gravitational waves in 2015 providing new insights into the behavior of the universe on the largest scales.

In the field of **metrology**, the science of measurement, quantum mechanics is transforming how we define fundamental units. Historically, measurements such as the kilogram or the second were based on physical objects or phenomena. For instance, the kilogram was once defined by a platinum-iridium cylinder stored in a vault in France. However, with the advent of quantum metrology, these definitions are being replaced by constants of nature that can be precisely measured using quantum techniques. In 2019, the definition of the kilogram was redefined based on the Planck constant, a fundamental quantum constant, using a device called a Kibble balance, which leverages quantum principles to measure mass with extraordinary accuracy. This shift to quantum-based standards ensures that measurements can remain consistent and accurate over time, regardless of the physical degradation or variability of traditional reference objects.

Quantum sensors are also showing potential in fields like biology and chemistry. For instance, quantum sensors can be used to detect extremely small variations in electric fields or changes in molecular structure, which could lead to more sensitive medical diagnostic tools or more efficient chemical analysis methods. In medical imaging, quantum-enhanced MRI machines could provide much higher resolution images, helping doctors detect diseases at earlier stages. Similarly, in drug discovery, quantum sensors might be able to monitor molecular interactions in real-time, accelerating the development of new treatments.

While quantum sensors are still in their early stages of development, the potential applications are vast. From improving the precision of GPS navigation systems to monitoring climate change by detecting minute variations in atmospheric conditions, quantum sensors have the potential to revolutionize a wide array of industries. They also have significant implications for the advancement of science itself. By allowing scientists to measure quantities that were previously too small or too subtle to detect, quantum sensors open up new possibilities for exploration and discovery.

However, developing practical quantum sensors is not without its challenges. Quantum systems are highly sensitive, not just to the quantities they are meant to measure, but also to environmental noise. Any unintended interaction with the surrounding environment can cause a quantum system to lose its coherence—a phenomenon known as **decoherence**—which can degrade the sensor's performance. Overcoming this obstacle requires advances in isolating and controlling quantum systems, a technical hurdle that researchers are actively working to address.

Quantum Mechanics and Artificial Intelligence

The intersection of quantum mechanics and artificial intelligence (AI) is an exciting and rapidly evolving area of research. At first glance, quantum mechanics and AI might seem like entirely separate fields—one focused on the fundamental laws governing the behavior of particles and forces at the smallest scales, the other on the development of intelligent algorithms and systems that can learn, reason, and make decisions. However, recent advances suggest that these two fields are increasingly converging, with quantum mechanics offering new possibilities for the development of AI, and AI providing powerful tools to help unlock some of the deepest mysteries of quantum systems.

Quantum mechanics is known for its counterintuitive principles, such as superposition and entanglement, which govern the behavior of particles at the atomic and subatomic level. These principles have already revolutionized our understanding of the physical world, and they are now beginning to influence the world of computation, particularly with the development of quantum computers. A quantum computer operates on qubits, which, unlike classical bits, can represent multiple states simultaneously due to superposition. This gives quantum computers the potential to process vast amounts of data much more efficiently than classical computers, especially for certain types of problems.

This is where artificial intelligence comes in. AI systems, particularly those based on machine learning, require enormous computational resources to process data, learn from it, and make predictions. Classical computers, though highly advanced, are limited in their ability to handle the most complex problems efficiently, such as optimizing huge data sets or simulating intricate systems like molecules or financial markets. Quantum computing offers a new way forward by potentially providing the computational power needed to solve these problems much faster than is currently possible.

The application of quantum mechanics to AI has led to the development of **quantum machine learning**, a subfield that explores how quantum computing can improve AI algorithms. By harnessing the power of quantum superposition and entanglement, quantum machine learning algorithms could process data in ways that classical algorithms cannot, potentially offering exponential speed-ups for tasks like pattern recognition, data classification, and optimization. For example, quantum algorithms could analyze vast data sets, identifying patterns and making predictions much faster than classical algorithms, which must process each piece of data sequentially. This speed-up could be particularly useful in fields like healthcare, where AI is used to sift through massive amounts of medical data to identify trends and recommend treatments.

Quantum mechanics also plays a role in the optimization of AI systems themselves. Many machine learning algorithms rely on optimization techniques to find the best solution to a problem, whether it's training a neural network or finding the shortest path in a logistics system. Some of these optimization problems are incredibly difficult for classical computers to solve, especially when they involve many variables or a highly complex landscape of possible solutions. Quantum computers, with their ability to explore multiple possibilities simultaneously, could offer new approaches to solving these problems more efficiently, allowing AI systems to learn faster and more effectively.

On the other side of this intersection, AI is also being used to advance our understanding of quantum mechanics. Quantum systems, particularly those involving many particles, are notoriously difficult to simulate on classical computers because the complexity of the system grows exponentially as more particles are added. AI algorithms, especially those that can learn and adapt over time, are being used to simulate quantum systems more efficiently. Machine learning techniques can help researchers model quantum behaviors, predict the outcomes of quantum experiments, and even discover new quantum phenomena that were previously too complex to study.

One of the most promising applications of AI in quantum mechanics is in the discovery of new materials and chemicals. In fields like quantum chemistry and materials science, researchers are interested in predicting how molecules will behave or how new materials will perform based on their atomic structure. These are quantum mechanical problems, as the behavior of atoms and electrons follows the laws of quantum mechanics. Simulating these systems using classical computers requires enormous computational resources and can take an impractically long time. AI, particularly machine learning models, can help by identifying patterns and making predictions about the behavior of these quantum systems, accelerating the discovery process. This could lead to the development of new materials with applications in energy, medicine, and technology.

AI also holds promise in advancing our understanding of quantum computing itself. Quantum computers are incredibly complex machines, and developing efficient quantum algorithms is a major challenge. AI can assist in the design of these algorithms, helping researchers explore different strategies for solving quantum problems and identifying which approaches are most likely to succeed. Machine learning models can analyze data from quantum experiments and simulations, learning from the results to guide the development of more efficient quantum algorithms.

Furthermore, AI may also help address one of the key challenges in quantum computing: error correction. Quantum systems are extremely sensitive to their environment, and even small disturbances can lead to errors in quantum computations. Designing effective error correction techniques is critical for building practical quantum computers, but it is a highly complex task. AI systems, which excel at recognizing patterns and optimizing systems, are being explored as tools to help develop and refine error correction methods, potentially bringing us closer to the realization of fault-tolerant quantum computers.

Despite the enormous potential at the intersection of quantum mechanics and AI, there are still many challenges to overcome. Quantum computers are still in the early stages of development, and building large-scale, reliable quantum machines will take time. Meanwhile, developing quantum algorithms that can outperform classical ones is a complex task that requires a deep understanding of both quantum mechanics and computer science. Likewise, integrating AI with quantum systems requires significant advances in both fields, as well as new insights into how these technologies can complement each other.

Recommended Resources

For those looking to delve deeper into quantum physics beyond this introduction, there are many excellent resources available. Quantum mechanics is a broad and complex field, and further study requires a strong foundation in both conceptual understanding and mathematical techniques. Below is a guide to recommended materials that will help you expand your knowledge, from beginner-friendly books to more advanced texts and online resources.

1. Introductory Books

If you are just starting your journey into quantum mechanics, it's important to build a solid foundation with books that explain the core concepts in an accessible way. Several books are designed specifically for readers with little to no background in physics or mathematics, offering a more conceptual approach.

- **"Quantum: A Guide for the Perplexed" by Jim Al-Khalili.** This book provides a clear and concise overview of the key concepts of quantum mechanics, explained in a way that doesn't require a deep understanding of mathematics. Al-Khalili's engaging style makes it an excellent starting point for beginners.

- **"In Search of Schrödinger's Cat" by John Gribbin.** Gribbin's classic work is an approachable introduction to quantum theory and its paradoxes. It covers the historical development of quantum mechanics and introduces readers to the strange and counterintuitive world of quantum phenomena.

- **"The Quantum World: Quantum Physics for Everyone" by Kenneth W. Ford**. Ford offers a non-technical exploration of quantum physics, explaining complex topics such as wave-particle duality, uncertainty, and quantum entanglement in an accessible manner. The book includes helpful analogies and examples to guide readers through difficult concepts.

2. **Intermediate Books**

Once you have a basic understanding of quantum mechanics, you may want to move on to more detailed studies. These books provide a deeper exploration of the subject, including some mathematics, but they are still approachable for those who aren't ready for full-on textbooks.

- **"Quantum Mechanics: The Theoretical Minimum" by Leonard Susskind and Art Friedman**. Part of the "Theoretical Minimum" series, this book provides a deeper look at the mathematics behind quantum mechanics while still maintaining a relatively accessible approach. It's a great bridge between popular science books and more advanced studies.

- **"Quantum Mechanics and Path Integrals" by Richard Feynman and Albert Hibbs**. Feynman's approach to quantum mechanics through path integrals provides a unique perspective on the subject. This book is more mathematically oriented, but Feynman's intuitive explanations make it a valuable resource for those looking to move beyond conceptual understanding.

3. Advanced Textbooks

For readers who are ready to dive into the full mathematical formalism of quantum mechanics, there are several comprehensive textbooks available. These are typically used in university courses and provide the rigorous foundation necessary for advanced study or research.

- **"Principles of Quantum Mechanics" by R. Shankar**. Shankar's textbook is a favorite among students for its clear explanations and comprehensive coverage of both the theory and mathematics of quantum mechanics. It starts with basic concepts and builds up to more complex topics, making it suitable for both undergraduate and graduate students.

- **"Introduction to Quantum Mechanics" by David J. Griffiths**. Griffiths' book is widely regarded as one of the best introductory textbooks on quantum mechanics. It balances clarity and rigor, making it an ideal choice for students who have some background in mathematics and classical physics but are new to quantum theory.

- **"Quantum Mechanics: Concepts and Applications" by Nouredine Zettili**. This textbook offers a detailed exploration of quantum mechanics with plenty of worked examples and exercises. It is an excellent resource for those looking to solidify their understanding of both the theoretical and practical aspects of quantum mechanics.

4. Online Courses and Lectures

The internet offers a wealth of free and paid resources for those who prefer a more interactive approach to learning. Many prestigious universities provide online courses that allow students to study quantum mechanics at their own pace.

- **"Quantum Mechanics" on MIT OpenCourseWare**. MIT offers a series of free, online lectures and course materials covering the fundamentals of quantum mechanics. These resources are suitable for those who have a strong background in mathematics and are ready to tackle the subject in a rigorous academic way.

- **"Quantum Mechanics and Quantum Computation" on Coursera**. This course, offered by the University of California, Berkeley, provides an introduction to both the basic principles of quantum mechanics and the emerging field of quantum computing. It's ideal for those looking to explore the practical applications of quantum theory.

- **"The Theoretical Minimum" video lectures by Leonard Susskind**. Based on his popular books, Leonard Susskind's series of lectures, available for free on YouTube and other platforms, is an excellent way to learn about quantum mechanics from one of the leading theoretical physicists of our time. His clear explanations make challenging topics more accessible.

5. Research Papers and Journals

For those interested in the cutting-edge developments in quantum mechanics, reading scientific papers and journals is essential. While these materials are more technical and require a strong background in the subject, they offer insights into the latest discoveries and theoretical advancements.

- **"Reviews of Modern Physics"** This journal publishes comprehensive review articles on topics in quantum mechanics and other areas of modern physics. It's a valuable resource for advanced readers who want to stay up to date with the latest research.

- **"Physical Review Letters"**. This prestigious journal covers the latest experimental and theoretical work in physics, including quantum mechanics. Reading papers from this journal can give you a sense of the current research directions in the field.

6. Quantum Computing Resources

With the rise of quantum computing, many resources are becoming available to help beginners understand this complex but promising field.

- **"Quantum Computing: A Gentle Introduction" by Eleanor G. Rieffel and Wolfgang H. Polak.** This book is designed for readers with no prior knowledge of quantum mechanics or computer science and provides a thorough introduction to the principles behind quantum computing.

- **IBM Quantum Experience.** IBM offers a free platform where users can experiment with real quantum computers and learn about quantum programming. It's an excellent resource for those interested in the practical applications of quantum theory in computing.

Afterword

Quantum mechanics has revolutionized our understanding of reality, revealing a universe that is far more complex, interconnected, and mysterious than our classical intuition suggests. We have seen how quantum theory challenges our deepest assumptions about the nature of space, time, and matter.

Quantum physics is still an evolving field. Many questions remain open, and as our knowledge grows, so too will our understanding of the universe. Whether we are scientists or simply curious minds, continuing to ask questions and seek deeper truths is at the heart of the scientific endeavor. Quantum mechanics, with all its paradoxes and beauty, reminds us that the universe is a place of endless discovery.

Thank you for walking this fascinating path!

Best regards,

Robert Meis.

www.ingramcontent.com/pod-product-compliance
Lightning Source LLC
Chambersburg PA
CBHW031429210526
45464CB00005B/2116